The Development of BRIC and tl
Advantage

Yao Ouyang

The Development of BRIC and the Large Country Advantage

格致出版社
Truth & Wisdom Press

Springer

Yao Ouyang
Big Country Economy Research Center
Hunan University of Commerce
Changsha, Hunan
China

ISBN 978-981-10-9215-2 ISBN 978-981-10-0633-3 (eBook)
DOI 10.1007/978-981-10-0633-3

Jointly published with Truth and Wisdom Press

Printed on acid-free paper

This Springer imprint is published by Springer Nature
The registered company is Springer Science+Business Media Singapore Pte Ltd.

Foreword of Publication

We have decided to publish the "Contemporary Economics Series," in order to present systematically an overview of the contemporary economics and its progress, summarize and tap the existing and potential results in the research of contemporary economics and demonstrate its new orientation of development.

The "Contemporary Economics Series" is a large and comprehensive series of academic theories on economics at high level. It consists of three subseries: (1) library of contemporary economics; (2) library of translated literatures of contemporary economics; and (3) series of teaching references of contemporary economics. In the field of study, this series not only aims at the latest results in all traditional disciplines of economics, but also pays more attention to the new achievements in the cutting-edge front, boundary and comprehensive sectors of economics; in selecting works, extensive contacts were made with scholars both at home and overseas, to collect works with profound academic foundation, new and unique views and high level of writing. All efforts will be made so that the "library" will attain the highest level in the current field of economics in China, the "library of translated literatures" will cover famous works of celebrities in contemporary economics; and the "series of teaching references" will mainly include general-purpose teaching materials in famous schools of higher learning in other countries.

This series aims at the modernization and international standardization of economics in China, and efforts will be made to gradually complete the transformation of the economics in China from a traditional one to a modern one in the scope, contents, research methods, and the analysis technologies in the near future. It is our eager hope that economists support our work by offering high-quality standardized works on economics to this series, as our common efforts to raise the research level of economics in China and have it take foot in the forest of economics of the world.

We are looking forward with our economists to the future of the economics in China.

Chen Xin

Foreword

This book puts forward questions from the perspective of the reasons of China's economic miracle and the rise of BRIC, develops the concept of the Comprehensive Advantages of Large Country (CAOLC), reveals the formation mechanism of the CAOLC, proposes an analytical framework of the CAOLC, analyzes the characteristics of economic development mode of Large Country and the strategic direction of economic development based on CAOLC, and builds the theory embryo of the CAOLC.

By "comprehensive advantages of Large Country," we refer to some special advantages stemming from the characteristic of "Large Country" itself and its "diversification," and they are stemming from the combination of various favorable resources of Large Countries. The principle of "comprehensive advantages of Large Country" enables us go beyond the debate between comparative advantage and competitive advantage, thereby examining the economic development advantages of Large Countries in broader areas from a systematic perspective.

In this book, "comprehensive advantages of Large Country" is mainly referred to the facts of China, India, United States, Russia, and Brazil, and firstly is based on the national conditions of China. China's ranking in the world in terms of comprehensive international competitiveness has been "rising with shocks," what kind of advantages allow China possess such competence? Clearly the answer is not only labor resources advantage, and more importantly, it should be "Large Country" advantage and "transformation" advantage: China is a Large Country with unbalanced economic and technological development, some regions and sectors have the advantages of low-cost labor resources and application technologies, while others possess the advantages of high-tech and high capital-intensive industries; China is experiencing a transitional economy, in which some regions and sectors have some advantages of developing countries, while others have the advantages of developed countries. All these different advantages can be positive factors to enhance the economic competitiveness of China, and therefore should be integrated into the source and power system of China's economic development.

In the study of "comprehensive advantages of Large Country" from the perspective of the general principle of Large Country's economic development, a "Magic Gourd Model" is developed to describe the main path of "comprehensive advantages of Large Country" formation in countries with Large Country characteristics. Specifically, they include many aspects: the advantage of deepening division of labor, cost-reducing advantages and pillar industries advantages stemmed from economy of scale; the synergy advantage of regional economy, industries and products originated from economic differences, the adaptive advantage in human capital, technology adaptation, and product suitability resulted from economic diversity, and the stability advantage in industries, products, and employment resulted from the economic integrity. The "comprehensive advantages of Large Country" mainly embody in the advantage of scale and deepening division of labor, the advantage of diversity and complementarity, the advantage of heterogeneity and adaptability, and the advantage of independence and stability because of the "big" scale of a country.

The economic growth factors of Large Countries mainly include natural resources, financial capital, human capital, technological progress, market potential, and foreign trade. All these factors can be used to describe the analytical framework of "comprehensive advantages of Large Country" in the sense that these factors in Large Countries are characterized by large scale, differences, heterogeneity and diversity, and they create advantages of scale and adaptability. However, why has Chinese economy with Large Country characteristics experienced periods of economic depression? One of the important constraints of economic development is the institution; old and rigid institution hindered China's economic growth, leading to the failure of the transformation from potential "comprehensive advantages of Large Country" to the real advantage. Reform and opening up and institutional transformation invigorated the Chinese economy and accelerated economical development of China. Institutional innovation exerts important influence on the formation of comprehensive advantages of Large Country and the optimization of various factors such as natural resources, human capital, financial capital, technological progress, market potential, and foreign trade, forming an environment conducive to economic aggregate and a suitable economic aggregate pattern, to provide motive force to economic aggregate and transformation. The successful experiences of reform and opening up in China are as following: adhering to institutional innovation with cultivating the market mechanism as the main line, to build a socialist market economic system, so that economic growth factors have been optimized and allocated rationally; adhering to the gradual reform path, so that institutional innovation in a Large Country made progress steadily; persisting in the combination of reform and opening up, so that a big economy can take full advantage of domestic market and international market, to drive economic growth with both domestic and international demands; adhering to transformation of economic growth by phase, so that a big economy could utilize latecomer advantage and first mover advantage, achieving the transformation and upgrade of industrial structure; and adhering to the development of appropriate technology and high-tech and the combination of imitative and self-dependent innovations, to develop various factors by technological progress.

The economic development mode of a Large Country has its own particularities, for which appropriate strategic direction should be set. For example, according to the economy scale of a Large Country, industry development should follow the strategy of overall advancement; according to economic development differences of a Large Country, regional development should follow the coordinated development strategy; according to the endogenous economy characteristic of a Large Country, both internal and external cyclic development strategy should be taken in the opening-up; and according to the independent economy characteristic of a Large Country, a proactive strategy should be adopted for economic development. The suitability of comparative advantage strategy should be investigated from the perspective of economic growth stage. Comparative advantage strategy is applicable mainly to the stage of quantity-oriented economic growth, instead of the stage of quality-oriented economic growth. During the transformation of growth pattern, the comparative advantage strategy should be re-examined. Specifically, the characteristics of China as a Large Country have made it possible to realize industrial structure upgrading more effectively. Given the multiple characteristics of a Large Country, the goals and strategy paths of different industries and technological upgrade in different regions should be different, instead of definitely sticking to the strategy of comparative advantage. At present, more in-depth discussion should be conducted in the selection of China's economic development strategy. The development strategy of China should reflect not only the huge effect of comparative advantage of labor resources on promoting employment and economic growth, but also the important role of science and technology in achieving breakthroughs in the technology and knowledge intensive industries.

The rapid rise of BRIC further highlights the big economy phenomena, and after the financial crisis the world economy sees the backbone role of U.S. as well as the newly rising Large Countries. BRIC should bring into full play their advantages as Large Countries, take an active part in establishing a new international financial system, and promote the sustainable development of world economy. Facing the huge impact caused by international financial crisis, China has taken positive actions to the resuscitation of global economy, by right of its advantages as a Large Country, through stimulating its domestic demand. In the post-crisis period, China should strengthen these actions, adjust and optimize its industrial structure, and improve the quality and effectiveness of its economic development, displaying a fully new figure to the world.

Yao Ouyang

Contents

About the Author

Yao Ouyang was born in Ningyuan County of Hunan Province in 1962. He studied in Lingling Branch of Hunan Normal College, Renmin University of China and Hunan University, and got the degrees of Master of Philosophy and Doctor of Economics. He became a professor in 2000. He is a vice president of Hunan Normal University and a Ph.D. supervisor. He was invited as a visiting research fellow by Hitotsubashi University of Japan in 2004, and as a senior visiting scholar by Stanford University in 2012, and worked as a visiting scholar in Georgia Institute of Technology, U.S. in 2007. He worked in the CPC Committee of Hunan Province and of Dong'an County and the People's Government of Yongzhou City, in 2006, he took office as Secretary of CPC committee and Director of Large Country Economy Research Center of Hunan University of Commerce, and concurrently as Vice President of Emerging Economics Research Association of China and Vice President of Commercial Economy Society of China. His main research orientation is international economics and development economics. He has taken charge of researches of a major project of National Social Science Foundation, a major project of National Soft Science Research Program, a project of National Natural Science Foundation, a project of Humanistic And Social Science Research Plan of the Ministry of Education, and a project of China Post-doctorate Science Foundation. He has published over 150 articles on newspapers and journals including Social Sciences in China, Economic Research Journal, *Management World, China Industrial Economy, Chinese Rural Economy* and *China Soft Science*, was awarded World Political Economy Outstanding Achievement Prize, An Zijie Prize of International Trade Research, first prize of social science research results of Hunan Province, selected to the First National Leading Talent "Ten thousands person plan" of philosophy and Social Sciences 全国 "万人计划" 第一批哲学社会科学领军人才 (译) "New century excellent talent support program of the Ministry of Education" and "New Century Talents Project of China," and receives special government allowance from the State Council.

About the Authors

Chapter 1
Introduction

1.1 Exploring Chinese Miracle

China has sustained rapid economic growth during reform and opening up. From 1994 to 2007, China's ranking in the world in terms of international competitive power rose from the 34th to the 20th place with concussions, and reaching the 15th place in 2007, which was the first place among transitional countries in overall level, leading far away some small developing countries. This has attracted world attention and is referred to as "Chinese Miracle". The Nobel Economics Prize laureate Joseph Stiglitz pointed out: "there has never been such large scale and lasting economic growth in the world. During the past 25 years, the annual average growth rate of China was 9 %, and the per capita income increased by 4 times (from 220 USD to 1100 USD)".[1] The growth of economic aggregate of China during 1978–2009 is as shown in Table 1.1.

It can be seen that during a period of 31 years, the annual economic growth rate of China basically maintained over 8 %, which is rarely seen in the history of world economic development. In recent years, scholars both at home and overseas have been exploring the reasons behind the sustained rapid growth of China's economy, and domestic scholars have put forth some different views in interpreting the "Chinese Miracle":

1. The view of "comparative advantage"
   It is represented by Justin Yifu Lin and Cai Fang, they believed that the key to the rapid economic growth in China is that the resources comparative advantage of China has been brought into full play by reforming the traditional economic system, and developing industries with intensive use of the factor of labor force. They advocated that developing countries should develop economy by following the comparative advantage strategy, and import advanced technologies from other countries with the accumulation of capital and changes of resource endowment to upgrade their industries, to push forward technological innovation in economy at low cost and maintain a long-term rapid growth.[2]

Y. Ouyang, *The Development of BRIC and the Large Country Advantage*,
DOI 10.1007/978-981-10-0633-3_1

Table 1.1 Growth of economic aggregate of China during 1978–2009

Year	GDP (100 m yuan)	Primary industry (100 m yuan)	Secondary industry (100 m yuan)	Tertiary industry (100 m yuan)	Per capita GDP (yuan)	Economic growth rate (%)
1978	3645.22	1018.40	1745.20	881.60	381.23	11.70
1980	4545.62	1359.40	2192.00	994.20	463.25	7.8
1985	9016.04	2541.60	3866.60	2607.80	857.82	13.50
1989	16,992.32	4228.00	7278.00	5486.30	1519.00	4.10
1990	18,667.82	5017.00	7717.40	5933.40	1644.47	3.80
1991	21,781.50	5288.60	9102.20	7390.70	1892.76	9.20
1995	60,793.73	12,020.01	28,679.46	20,094.30	5045.73	10.90
2000	99,214.55	11,716.22	45,555.88	38,942.50	7857.68	8.40
2001	109,655.20	15,516.17	49,512.29	44,626.70	8621.71	8.30
2002	120,332.70	16,238.62	53,896.77	50,197.30	9398.05	9.10
2003	135,822.80	17,068.32	62,436.31	56,318.10	10,541.97	10.00
2004	159,878.30	20,955.83	73,904.31	65,018.20	12,335.58	10.10
2005	185,808.60	22,420.00	87,598.09	74,919.28	14,185.36	10.40
2006	217,522.70	24,040.00	103,719.50	88,554.88	16,499.70	10.70
2007	267,763.70	28,627.00	125,831.40	111,351.90	20,169.46	11.40
2008	316,228.80	33,702.00	149,003.40	131,340.00	23,707.71	9.00
2009	343,464.70	35,226.00	157,638.80	147,642.10	25,575.48	8.70

Source China Statistical Yearbook (2010)

2. The view of "later-mover advantage"
   It is represented by Guo Xibao and Shi Donghui, they held that the rapid economic growth of China was mainly sourced from a "later-mover advantage", which is paragenetic with the relatively backward economy in late-developing countries, and is an advantage derived from their existing backwardness. Guo Xibao proposed a framework of five later-mover advantages including capital, technology, manpower, institutional system and structure.[3] Shi Donghui analyzed the strategic schema on industrialization in late-developing countries.[4] They advocated that developing countries implement a catch-up strategy based on "later-mover advantage", i.e., they can learn the successful experience and lessons of failure of developed countries, to gradually narrow the gap to developed countries on the basis of imitative innovation.
3. The view of "Large Country advantage"
   It is represented by Hai Wen and Li Daokui, and they believed that the sustained economic growth in China is mainly backed by the "advantages of Large Countries". Hai Wen believed that China has an "advantage of Large Country" different from that of other Eastern Asia countries, including the long-term comparative advantage of labor force, huge market size, the space of institutional reform, and the favorable factors of national environment.[5] Li Daokui believed that China's economic development has an urgent demand for

exploring on Large Country strategy, if Large Countries only keep on pursuing for comparative advantage, they will come to an awkward situation that "what they make is of low value while what they don't make is of high value", therefore, they should not stay with the traditional comparative advantage strategy, instead, they should actively implement a Large Country strategy, to get initiative in international economic competition.[6]

All the above views are reasonable to a certain extent, explaining the causes of sustained rapid economic growth of China from different angles. Relatively speaking, "advantages of Large Country" is a comparatively complete and reasonable explanation, however, systematic research is lacking on the "advantages of Large Country" in the academic circle today. Usually only some factors are listed, without obtaining well-organized integral results. In the view of the author, the key to "advantages of Large Country" lies in its "comprehensiveness", therefore the concept of "comprehensive advantages of Large Country" (CAOLC) is proposed, with an attempt to make a scientific and reasonable explanation to the sources of the economic competitiveness of China. The CAOLC proposed on the basis of the special national conditions of China is mainly demonstrated by the advantages of "Large Country" and "transformation": the unbalanced economic and technological development in China has resulted that some areas or sectors have the advantages of labor resources and suitable technologies, and some other areas or sectors have the advantages of capital and high and new technologies; China is in the period of economic transformation, and the progress in different areas and sectors transformation is not balanced either, some areas or sectors have the advantages of developing countries, and some other areas or sectors have the advantages of developed countries. By optimizing and combining these different advantages in a scientific and rational way, we can form a collection of active factors to enhance the economic competing power of a country, that is the "comprehensive advantages of Large Country".

1.2 Rise of BRIC

With the rapid rise of BRIC, the phenomenon of "Large Country economy" is becoming all the more attractive in the world. China, India, Russia and Brazil, as main emerging market economies, their gross domestic product (GDP) is now accounting for 14.6 % of the world total. Their trade volume accounts for 12.8 % of global total, and their contribution to world economic growth has exceeded 50 % when calculated by purchasing power parity (PPP). According to statistics by the International Monetary Fund (IMF), from 2006 to 2008, the average economic growth of the four countries was 10.7 %, ranking at the front among all countries in the world. This is a new phenomenon in international economy, and after the "East Asia Miracle" made by the "four little tigers" in Asia, the BRIC are making a new "miracle in the world" (Tables 1.2 and 1.3).

Table 1.2 GDP growth rate in BRIC countries in **1990–2008** (%)

Country	Year									
	1990	2000	2001	2002	2003	2004	2005	2006	2007	2008
China	3.83	8.42	8.30	9.09	10.02	10.08	10.43	11.60	13.00	9.00
India	5.81	4.03	5.22	3.77	8.37	8.30	9.33	9.67	9.05	6.00
Russia	N.A.	10.00	5.10	4.70	7.30	7.20	6.10	7.70	8.10	5.60
Brazil	−8.40	4.28	1.31	2.67	1.17	5.70	3.15	3.94	5.65	5.10

Source According to the statistical yearbooks of respective BRIC countries

Table 1.3 Proportion of total GDP of BRIC countries in the world in **2000–2007** (%)

Country	Year							
	2000	2001	2002	2003	2004	2005	2006	2007
China	7.20	7.70	8.10	8.70	9.10	9.60	10.20	10.80
India	3.70	3.70	3.80	3.90	4.00	4.20	4.40	4.60
Russia	2.70	2.80	2.80	2.90	3.00	3.10	3.10	3.20
Brazil	3.00	2.90	2.90	2.90	2.90	2.90	2.80	2.80

Source IMF world economic outlook database 2008 Apr

What are the common causes behind the rise of the BRIC? These countries have different social systems and ideologies, and they also have many differences in the transition of economic system and economic development level, therefore a reasonable explanation cannot be obtained in these aspects.

Some scholars tried to analyze the causes in "strategic pattern": these countries have adopted the comparative advantage strategy, which enabled full play of their natural resources, to push the growth of economy at high speed over a long period at low cost and with good efficiency. However, let alone that the strategic patterns of the four countries in economic development are not fully identical, why so many countries in the world adopting comparative advantage strategy have not realized rapid growth?

Some scholars tried to analyze the causes from the viewpoint of "later-mover advantage", on the ground that the previous relatively backward economy in these countries could form an advantage for development. This could provide large space for long-term rapid economic growth. However, there are many other countries with relative backward economy in the world, besides the BRIC, and they should also be able to form and make use such later-mover advantage, to push forward rapid economic growth by improving their technologies and human resources, economic structure and institutional system of their own. Then why economic development in some of these countries has remained slow, without the phenomena of rapid growth?

Therefore, all these explanations above are not quite convincing. In the view of the author, two basic principles should be followed in search of the causes of the rise of BRIC: first, the cause should be common to all four countries but not available in most other countries; and second, this cause has direct and inevitable link with economic growth. Then we can see clearly that the BRIC countries are all newly rising Large Countries, with the characteristics of Large Countries: their "big" territorial area, population size, total resources and market potential have provided favorable conditions for economic growth. China, India, Russia and Brazil rank respectively at the 4th, 7th, 1st and 5th places in the world in terms of land area, at the 1st, 2nd, 7th and 5th places in the world in population size, at the 3rd, 8th, 2nd and 9th places in the world in gross resources, and at the 8th, 12th, 16th and 7th places in the world in market scale. It is the "Large Country effect" generated by the characteristics of Large Country economy that formed the advantages in market, resources, scale as well as on diversity, complementarity, adaptability, pushing forward the sustained rapid growth of economy in the BRIC.

1.3 Research Schema and Theoretical Contributions

1.3.1 Objects of Research

Just as the name implies, the objects of research of the "comprehensive advantages of Large Country" theory are surely "Large Countries". However, there is no unified criterion for the definition of a "Large Country". For example, the American economist Simon Kuznets' classification was mainly based on population size, while Almond and Junior Powell divided countries into "Large Country" and "small country" according to territorial area and population size. Chenery made his classification using a number of indicators including population size and trade index, and also adjusted his classification criterion a number of times.

Zhang Peigang, the first person in China engaged in the research of development economics, did special research on "developing Large Countries", and he believed that "developing Large Country" is a comprehensive concept incorporating both natural and geographic characteristics as well as social and economic aspects. Specifically, they are developing countries with large population, vast territory, rich resources, long history, and low per capita income level. Zhang Peigang also summarized the empirical characteristics of the development of late-developing Large Country economies: (1) the domestic demand and market capacity are comparatively big; (2) the large-scale infrastructures and domestic demand need large amount of capital; and (3) in the starting stage of industrialization, an economic system with a complete range should be established.

The author basically agrees with the classification criteria of Zhang Peigang, but on the ground that long history is not surely linked with a Large Country, and the low per capita income level is a feature of late-developing Large Countries, proposes to take the first three factors, plus market size, as the basic criteria to classify "Large Country" and "small country": (1) vast territory; (2) large population; (3) rich resources; and (4) market potentials. Out of the need in studying the

CAOLC, the author will select five extra-Large Countries as the objects of research, and they are China, India, United States, Russia and Brazil.

To define the objects of research, it will be necessary to distinguish “Large Country” from “powerful country”, and late developing Large Countries from developed Large Countries. First, “Large Country” and “powerful country” have the difference of comprehensive category and economic one. A “Large Country” is defined by natural factors, social factors and economic factors, for example, vast territory, rich resources are natural factors, large population is a social factor, and market potential is an economic factor; while a “powerful country” is defined mainly by economic factors. The concepts “Large Country” and “powerful country” are often confused in some researches in China, while the author studies CAOLC mainly to explore the mechanism of an “ordinary Large Country” becoming a “Large Country of economy”, or to explore the mechanism of a “Large Country” becoming a “powerful country”, therefore “comprehensive Large Country” should be taken as object of research. Secondly, late developing Large Countries and developed Large Countries are different only in level within “Large Countries”. Late developing Large Countries are in the primary stage of a “Large Country” becoming a “powerful country”, while developed Large Countries are in the advanced stage of a “Large Country” becoming a “powerful country”. The author will start his research from late-developing Large Countries, and then expand it to developed Large Countries, and some common general characteristics will be summarized through comparison and studies.

China is a late developing Large Country, and is also the first object of our research. In the opinions of the author, in studying the comprehensive advantages of late developing Large Countries represented by China, the following two issues should be well steered: (1) the basic characteristics and staged characteristics. The former exists in the whole course of development, for example, the development of regions or sectors in a Large Country is usually unbalanced; and the latter demonstrates only in specific stages, for example, there must be different characteristics in the primary stage and advanced stage. (2) Long-term advantages and short-term advantages. For example, the labor resource advantages in China have the different characteristics of short-term and long-term, in short term, China has labor resources with low cost and low quality; and in long-term, there will be labor resources with fairly low cost and fairly high quality.

1.3.2 Contents of Research

In this book, the method of systematic analysis will be used, to study the basic concept, formation mechanism, analysis framework and development strategy concerning CAOLC, with the attempt to build a set of fairly complete theory on comprehensive advantages of Large Country.

(1) Basic concept. In Chap. 3 of this book, the evaluation indicator system for Large Countries is proposed starting from defining Large Countries, to analyze the levels and types of Large Countries; then the characteristics of Large Country economy will be expounded, to reveal the main performance of Large Country advantages; and finally, the concept of CAOLC is stated, with exposition on the connotation and basic performance of CAOLC, so as to interpret the comprehensive advantages of Large Country and the economic competitiveness of China in a fairly rational manner.

(2) Formation mechanism. In this chapter, the formation mechanism of CAOLC is specially studied, and a theoretical model is built on this basis, by interpreting this theoretical model, the mechanism of CAOLC to promote industrial competitiveness, enterprise competitiveness and foreign trade competitiveness is analyzed; by analyzing the scale and division of labor, heterogeneity and adaptability, difference and complementarity, independence and stability of Large Country economy, the mechanism of forming the CAOLC is expounded; and finally, the mechanism of CAOLC converting into competition advantages in economic development is expounded.

(3) Analysis framework. In Chap. 5 of this book, the basic analysis framework for CAOLC is proposed by explaining the basic factors of economic growth. On this basis, the advantages of natural resources, financial capital, human capital, technological progress as well as product export are explained respectively. Finally, analysis and exposition are made on the mechanism of institutional innovation in late developing Large Countries promoting economic growth.

(4) Development strategy. Chapter 6 of the book is about the research of the economic development strategies of Large Countries, it will start from expounding the main characteristics of the development pattern of Large Country economy, to analyze the adaptability of comparative advantage strategy in the development stages of Large Country economy, and also reveal its inadaptability in the stage of transforming economic growth pattern in a Large Country. In the meantime, the economic development strategy orientation based on CAOLC is analyzed, and recommended policies are proposed on economic development in Large Countries.

(5) Case analysis, Chaps. 7, 8 and 9 of this book present specific cases for studying the development of Large Country economy, the adaptability of relevant factors in Large Country economy growth are taken as the main line, first, the heterogeneity of human capital in late developing Large Countries is analyzed, to reveal the adaptability of different human capital to different material capital and technological level; then the difference in the development of different regions in late-developing Large Countries is analyzed, to reveal the mechanism how the complementarity and adaptability produced by such difference promotes the coordinated development of regional economy in a Large Country; and finally, the diversity of out-bound direct investment of late-developing Large Countries is analyzed, to reveal the adaptability in different industries, enterprises and regions.

1.3.3 Research Methods

Marx wrote in the preface of the original version of *The Capital*: "To analyze the forms of economy, only abstract logic thinking can be used".[11] To analyze the social and economic phenomena correctly, clear concept, proper judgment and logic reasoning must be ensured. The materialistic dialectics is a scientific logical thinking, and proceeding from concrete to abstract, and then from abstract to concrete is a basic approach in dialectic thinking. Any actual and complete cognition is a process from sensible concrete to abstract, and then ascending from abstract to concrete in thinking. In the research of CAOLC, we should follow this inevitable process of cognition, first, the concept of CAOLC should be put forth by analyzing the Large Country effect in economic growth in the economic characteristics of Large Countries; then the concept of CAOLC should be used as a logic start point, to make more clear and accurate the theory of CAOLC in concrete thinking through analyzing the theoretical model, basic framework, strategic pattern as well as relevant cases of CAOLC.

Scientific logical thinking is an organic whole, and various scientific approaches should be applied to complete the process from concrete to abstract, and then from abstract to concrete. For example, by applying the approach of combining analysis with synthesis, the analyses of particularities can be synthesized into general conclusions, that is, the analysis of Large Country effect of China, India, Russia, Brazil and United States can be summarized as CAOLC, and things of general nature can be further decomposed into various elements for in-depth analysis, that is, all relevant factors in the Large Country economic growth can be analyzed in detail, and the method of comparison and analysis can be used, to compare the Large Country effect in the economic growth in China, India, Russia, Brazil and United States, to extract the common natures of all Large Countries; and comparison can also be made on the effect of economic growth in Large Countries and small countries, to extract the individual characteristics of Large Country economy different from that of small countries.

Mr. Sun Yefang said: "In studying economic issues, we must distinguish the special experience and general laws of China or other countries". "Some special experience of China is also of general significance to countries with identical or similar social and historical conditions of China".[11] Therefore, we can use the approach of case analysis, that is, taking China as a case to analyze the characteristics of China as a Large Country and in transition, to summarize the comprehensive advantages of economic growth in late-developing Large Countries, so as to give play to the general significance and reference function of China's experience to late-developing Large Countries.

In scientific research, determining the nature of things is qualitative research, and analysis of quantities of things is quantitative research, and the latter can make the former more precise. Therefore, in studying the Large Country economy and the CAOLC, we should well combine qualitative analysis with quantitative analysis, make empirical analysis of all characteristics and effects in the economic growth in Large Countries by using the methods of statistic prevention and modeling analysis, to make the research of CAOLC more precise.

1.3.4 Theoretical Contributions

The theoretical contributions of this book are mainly on three aspects:

First, it put forth the concept of "comprehensive advantages of Large Country", and made new interpretation to its connotation. Large Country economy is a special economic phenomenon, and the theory of comprehensive advantages of Large Country is about the characteristics, advantages and strategic selection in the development of Large Country economy. It is a special advantage originating from the scale, difference and relative independence of Large Country economy, and can integrate comparative advantage and competitive advantage. This concept can explain in a fairly rational way the source of China's economy competitiveness as well as the Large Country economy phenomena in the rise of "BRIC".

Second, it has revealed the formation mechanism of comprehensive advantages of Large Country, and analyzed the basic route of its formation. The book has analyzed the formation mechanism of CAOLC in scale, difference and relative independence, revealing the economic development advantages of scale effect, adaptability and relative stability derived from it. Especially, the analysis of difference and complementarity advantages, diversity and adaptability advantages has provided a new angle of view to reveal the economic development advantages of late-developing Large Countries.

Third, it has expounded the economic development strategy based on CAOLC, and analyzed the suitability of comparative advantage strategy. The book has put forth the question of adaptability of comparative advantage strategy in different stages of development of Large Country economy, with a view that this strategy is adaptable to the primary stage of Large Country economy development, but not suitable to the new stage of economic growth pattern transition. It has made breakthrough to the view that developing countries can only adopt the comparative advantage strategy, and put forth the idea of combining comparative advantage with competitive advantage strategy. In the meantime, it has expounded the economic development strategy orientation based on CAOLC, and made proposal on relevant policies.

Notes

1. Quoted from: Liu Xiahui, Zhang Ping, Zhang Xiaojing: *Economic Growth and Structural Transformation in Era of Reform*, Truth & Wisdom Press, Shanghai People's Publishing House, 2008 Edition, page 3.

2. Prof. Justin Yifu Lin studied the miracle of China's economic reform and development by using the modern economy analysis method, and believed that late-developing countries cannot adopt catch-up strategy, and can be successful only by implementing the comparative advantage strategy, refer to: Justin Yifu Lin, Cai Fang, Li Zhou: *China's Miracle: Development Strategy and Economic Reform (Revised and Enlarged Edition)*, Shanghai Sanlian Bookstore, Shanghai People's Publishing House 1990 Edition.
3. Guo Xibao, Hu Hanchang: *New Theory of Later-mover Advantages*, *Journal of Wuhan University* 2004 Vol. 3, Guo Xibao, Hu Hanchang: *Later-mover Advantage Strategy and Comparative Advantage Strategy*, *Jianghan Tribune* 2003 Vol. 9.
4. Shi Donghui: *Introduction to Industrialization in Late-mover Countries: A Survey of History of Industrialization and Industrialization History*, Shanghai University of Finance and Economics Publishing House 1999 Edition.
5. Hai Wen: *China's Economy Has Advantages of Large Countries and Can Grow Powerfully for Further 20 Years*, *Beijing Morning Post* June 1, 2007.
6. Li Daokui *China's Economy Needs Large Country Development Strategy*, *China Today Forum* 2006 Vol. 7.
7. Data from IMF net.
8. Gross resources are expressed by national mineral reserves, market size by household final consumption expenditure, and data are taken from *China Statistical Yearbook* (2002) and *China Statistical Yearbook* (2006).
9. Zhang Peigang: *New Development Economics*. Henan People's Publishing House 1992 Edition.
10. Compiled by Central Compilation & Translation Bureau of Works of Marx, Engels, Lenin and Stalin: *Complete Works of Marx and Engels* People's Publishing House 1997 Edition.
11. Mr. Sun Yefang is the first expert stating that the special experience of China would have general reference significance to countries with similar conditions, and he predicted the world-wide significance of the China model. Refer to: Sun Yefang: *Manuscript of Socialist Economic Theory*, Guangdong Economic Press 1998 Edition, page 7.

Chapter 2
Related Literatures on Large Country Economy

In Large Countries with vast national territorial area, large population, rich resources and massive market, the economic development has characteristics different from that of small country economy, especially, the attractive resource advantages and huge domestic market have determined their unique track of economic development. Therefore, scholars both at home and overseas started long ago paying attention on the effect of country scale on economic development, conducted some explorations on country scale and Large Country economy, and put forth some views on development of Large Country economy.

2.1 The Evolution of Foreign Literature

2.1.1 Theoretical Origin

Research by economists on Large Country economy can be dated back to the earliest ancestor of economics Adam Smith. In his monumental work *An Inquiry into the Nature and Causes of the Wealth of Nations*, he explained the idea that "division of labor is restricted by the scope of market". "Division of labor originated from the exchange capacity, and the extent of division of labor, is therefore restricted by the exchange capacity, or in other words, by the broadness of market".[1] A market too small in size cannot encourage people to be engaged in a special work all their lives, therefore many occupations can be operated only in big cities, as broad market can provide good conditions for division of labor and concentration of industries. The feature of economic scale in Large Countries can make the division of labor quite developed both in regions and in countries, and this idea can be regarded as the theoretical origin of scale economy and Large Country economy.

Y. Ouyang, *The Development of BRIC and the Large Country Advantage*,
DOI 10.1007/978-981-10-0633-3_2

The *Principles of Economics* of Alfred Marshall provided nutrient to the theory of Large Country economy mainly on two aspects: one is that it expounded the completeness and complementarity of economy in big industrial areas. In small industrial areas mainly depending on a single industry, economy can become extremely depressed due to reduced demand or raw materials supply; however, big industrial areas with many different types of well developed industries can avoid such a problem. Second, it expounded the scale effect of large-scale production. "The main interest of large-scale production is an economy of technologies, an economy of machinery and an economy of raw materials".[2]

The *Theory of Economic Growth* by Arthur Lewis summarized systematically the factors influencing the economic growth, providing an analysis framework for the economic growth in Large Countries. Just as Lewis said, "I want to draw a blueprint instead of setting a theory".[3] This blueprint covers many factors associated with economic growth, such as economic system, market size, knowledge accumulation, capital accumulation, population and resources. To study the growth of Large Country economy, a blueprint should also be drawn by following such a framework.

In his work *Scale and Scope: the Dynamics of Industrial Capitalism*, Alfred Chandler Jr. specifically analyzed the economies of scale and economies of scope in production. He believed that it was the development of new technologies and opening of new market that led to economies of scale and economies of scope, and reduced transaction cost. "The utilization of economies of scale in production and marketing, the utilization of economy of joint production or joint marketing, or the reduction of relevant transaction cost, have cut cost and led to high efficient utilization of resources".[4] The economies of scale and economies of scope in production are mainly demonstrated by reduced unit cost; and the economies of scale and economies of scope in marketing are mainly manifested by the cost advantage of middlemen. Producers and distributors increase the cost advantages by expanding scale and scope, and this is economies of scale and scale efficiency. Large Countries have vast regional space and market space, and can also use their economies of scale and economies of scope to form advantages of Large Country.

2.1.2 Pattern of Large Countries

In the 1960s, in the research of international statistics and comparison, economists started paying attention to the influence of country differences on economic activities. In his book *Modern Economic Growth*, Simon Kuznets made it clear that country was taken as a unit in the study of economic growth, and conducted preliminary exploration on Large Country economy: first, he talked about the concepts of "Large Country" and "small country", stating that some very small countries are satellite countries of a Large Country or Large Country group to a large extent, and some superpowers are formed by regions with entirely different economic backgrounds in the early period of economic development, therefore these points should be taken into consideration when a country is taken as a unit in the study of economic growth. Second, he mentioned the internal difference in the economic development in Large Countries, with a view that the analysis of economic structure difference and variations of countries should be linked with the analysis of the regional difference and degree of unification in the economic development stages in Large Countries. Third, he analyzed the dependence of Large Countries and small countries on international trade, concluding that small countries depend more on international trade, and small countries can achieve economic growth only with increased dependence on foreign trade, while Large Countries can achieve economic growth even with a very low proportion of foreign trade, and fourth, he analyzed the advantages of Large Country economy, with a view that in Large Countries, the "domestic market and resources conditions allow them to develop specialization and economies of scale", if small countries realize the advantages of specialization and economies of scale, they will naturally depend on international trade with Large Countries.[5]

In his book *Economic Growth of Nations*, Kuznets conducted further research on the "effect of size of countries" and analyzed the relationship between the size of country with foreign trade, scale of economy, domestic production structure and economic development level. In his view, "when the size of a country has major effect on the proportion of foreign trade in gross value of production, it produces influence of overseas supply and demand on domestic production value structure".[6] To analyze the per capita output value level of different countries, he used population as a scale with 10 million as a demarcation line, to divide countries into big (each with population of over 10 million) and small (each with population below 10 million) ones, and conducted empirical research on issues such as the difference of per capita output value of Large Countries and small countries and the influence of the size of a country on industrial institutions.

In the book *The Patterns of Development*, Hollis B. Chenery and Moises Syrquin specifically studied the forms of Large Countries and their scale effect, and they stated that, "the most obvious effect of large scale is reflected in production type. Corresponding to a fairly enclosed trade pattern with low degree of specialization is a fairly balanced type of domestic production with little variation, in addition, except for few exceptions, Large Countries adopted the policy of inward development, which has produced effect on other aspects of accumulation and resources allocation".[7] The main effect of large scale and low export on resources allocation is that it requires these countries to change the economic structure in fairly early stage of development; on accumulation, the investment and saving levels in Large Countries are high, therefore the economy can grow slightly faster. In the meantime, "Large Countries always focus on the issue of internal development, resulting in a whole set of introverting policies with extensive consequences".[8] Because of offsetting by comparatively great difficulties in politics and administration, Large Countries cannot create obvious advantages in the course of development.

Chenery and Syrquin specially analyzed the Large Country pattern in *Industrialization and Growth: A Comparative Study*, and they pointed out three patterns of state trade, i.e., Large Country pattern, finished product export oriented small country pattern and primary product export oriented small country pattern. In their view, "a Large Country tends to select an introverting policy, which seems more suitable to a Large Country than to a small country".[9] And it was also found that a Large Country reaches the quasi-industrial stage at about 250 USD, but it is about 400 USD with standard pattern and about 600 USD for a small country exporting primary commodities. "For the Large Country pattern, early start of industrialization can be attributed to the general adoption of import substituting policy in the manufacturing industry".[10] Chenery et la also studied the criteria for a Large Country, by setting the population size as 5 million for a small country, as 20 million for a Large Country, and as 10 million for a country as comprehensive sample. In 1989, he proposed a new criterion, classifying countries with a population between 50 and 150 million in 1980 as Large Countries, and those with population less than 5 million as small countries.

2.1.3 Characteristics of Large Countries

Large Country economy has its own characteristics, and diversification of economy and differences in resources are root causes leading to the difference in the development strategies of Large Countries.

European Union (EU) is a union of countries, and as an integral whole, it has most characteristics similar to the development of a Large Country. Therefore, Bernhardsen (2000) took the EU as a whole and Germany as objects of research, by taking panel data of 12 months to study the relationship between interest rate difference and macro economic variables. It was found in the study that the regression coefficient is substantial to the prediction result, especially, the inflation rate, actual income growth, relative labor cost and current capital accounts have fairly great influence on interest rate difference. Therefore, the macroeconomic policies of various governments influence the interest rate difference to a certain extent. This result is fairly optimistic as compared with earlier studies, and the 1-month and 3-month maturity interest rate can usually embody the variation of interest rate. Although overnight rate should best embody the interest rate difference, it is usually not so stable as it is often manipulated by central banks. If the overnight rate is set at an unbalanced level, market participants have the reason to expect that such an unbalanced interest rate is temporary. In this case, the 1–3-month interest rate will be able to better reflect the fundamentals of economy, which is in line with the results of early studies, the long-term interest rate often fluctuate around the actual interest rate, without too much deviation; this research result has important guiding significance to the economic development in Large Countries, that is, governments of Large Countries and local governments in Large Countries all do their best to avoid frequent change of economic policies and ensure stable fiscal and monetary policies, so as to ensure the stability of economic policies and predicted economic movement in Large Countries.

Investment is important in driving the economic development, and is a factor that must be paid attention to in the research of Large Country economy, however, traditional researches usually focused on the relationship between inventory investment and investment variation. Chikan et al. (2005) overcame the insufficiency of traditional researches and studied the macroeconomic conditions of 14 Large Countries in the world by using the OECD database in the view point of inventory investment, thereby summarizing the macroeconomic characteristics of Large Countries. They first made 7 assumptions, and verified them one by one, and the final verification results showed that, first, in rapid economic development period, the inventory investment is high; second, inventory investment and fixed asset investment are positively related; third, inventory investment and foreign trade are negatively related, and this also implies that inventory change and foreign trade are two substituting tools in macroeconomic regulation. These research results can help us analyze and predict the macroscopic behavior and policy trend of Large Country economies (for example, to the policy-makers of Large Country economy, large-scale investment is required to promote the rapid development of economy). The analysis above only suits short-term analysis, and this conclusion may not be applicable in long term, so it is required that Large Countries follow their own general laws and pay attention to exploration of long-term trend.

Large Country economy is defined and studies macroscopically, and a core in the study of macroeconomic is the fiscal policy. Jin and Zou (2005) made an in-depth research of the financial decentralization, revenue and transfer payment in a big developing country—China. In their view, traditional theoretical studies have proved that the matching between revenue level and transfer payment can allow sectors at all levels in the country to share interests and efficiency, therefore more effectively promote economic growth in a Large Country. However, research about China showed that at national level, it is diffusing instead converging between revenue level and transfer payment; at provincial level in China, the transfer payment and revenue level of government is converging, with the root cause of financial decentralization. A group of data of 30 provinces was used to check the relationship between financial decentralization and economic growth in China in two phases (the financial contract system during 1979–1993 and the system of tax distribution during 1994–1999) in two phases, and it was found that the contradictions between theory and reality can be conciliated in China, and the key is to adopt the effective institutional arrangement of financial decentralization. Development of Large Countries has obvious characteristics of diversity, and this requires Large Countries to take into full account the important bridge role of financial decentralization between revenue and transfer payment in the course of economic development, to promote the effective growth and coordinated development of regional economy.

Major economies are benchmarks of global economy, the economic movement and development strategies of Large Countries play an important leading role in global economy, and to keep known the evolution of major economies, macroeconomic prediction for Large Countries must be strengthened. Fildes and Stekler (2002) found in their research that the macroeconomic prediction for Large Countries was extensively used in the cognition of industrial development and economic cycles, it is an important basis for setting Large Country economic development strategy, and more important, the key to grasp the economic fluctuation in Large Countries. They made their researches with the United States and United Kingdom, the two countries with high proportion of economy, as examples, to predict the economy status of these two Large Countries by using substitutable time sequence data and fairly accurate macroeconomic prediction model, and it was found that to the United States, the error rate of using annual data is at least over 25 %, and it would be more effective to use quarterly data, enabling more effective prediction of the relationship between the economic growth rate and inflation rate of the United States, and the research of the UK also supported this viewpoint. Therefore, they believed that in the economic prediction of Large Countries, it would be more scientific and reasonable to use quarterly data, which can more effectively reflect the trend of world economic development.

Daily close economic ties have made the economic fluctuation of Large Countries more likely to conduct to other countries, especially, the economic fluctuation in the United States can produce important impact on all countries in the world. Ireland and Schuh (2008) used the two-phase true economic cycle model and the data of the United States after the World War II to study the relationship between productivity and macroeconomic performance of Large Countries, to determine the impact from different consumptions, investments and commodity production technologies to total factor productivity. It was found in the research that in the 1970s, the slowdown of total factor productivity was caused by lagging of consumption, but that impact was temporary, while the lagging of investment has lasting impact on the total factor productivity, the substantial rise of total factor productivity in the 1990s was an investment impulsion resulted from technological transformation, and it will be quite difficult for the total factor productivity development level in that period to occur again in a period of time in the future. This requires that Large Countries with important global influence should be good at grasping the opportunities of technological transformation, and make full use of the investment drive brought about by technological transformation and rise of total factor productivity to stimulate the development of global economy.

2.1.4 Rise and Fall of Large Countries

At the end of the 20th century, foreign scholars started thinking more about the rise and fall of Large Countries, and the American historian Paul Kennedy published *The Rise and Fall of the Great Powers* (2006). According to Paul Kennedy, the basic features of the rise and fall of great powers are the continual change of relative positions of great powers in world affairs mainly for two causes: the first is the different growing rate of national strength of different countries, and the second is technical breakthroughs and transformation of organizational forms, which can give more advantages to a country than to another. "We have found a causal relationship between the changes of relative comprehensive economic strength and production capacity and the positions of all great powers in the international system". Starting from the 16th century, the world trade concentration places gradually shifted from the Mediterranean to the Atlantic and Western and Northern Europe, and in the decades after 1890, the world concentrated producing places of industrial products shifted slowly from Western Europe to other regions, he stated with this example that, "the transfer of economic strength heralds the rise of new great powers". In addition to economic causes, many factors such as geographic location, military organizations, national morale and union system can govern the national strength of countries. He also grouped the Large Countries in the world today into five systems: the United States, Soviet Union, China, Japan and countries of European Economic Community.

The American scholar Schweitzer (2007) studied the scientific and technological and economic issues of Russia. He believed that, Russia possesses rich technical resources, and the Russian economy will rise again by promoting technological progress, "No matter how dim the economic prospects will be, the leaders in Moscow and other places will insist on their objectives, to make Russia a powerful industrial country again". German scholar Karl Pilny studied the influence of India and China on the world. He pointed out in his book *How India and China Change the World* (2008): the economic growth rate of these two countries with the largest population in the world has far exceeded the world economic growth rate, and he predicted that they will actively remain as the world economic growth centers, rising to become superpowers in the 21st century. Pilny analyzed the common characteristics of the two countries: a huge population, and with a large rural population; the economic structure is dominated by agriculture; and the economic system is in the transformation process. At the same time, he analyzed the different development patterns of the two countries: the Indian pattern is mainly founded upon the service industry; but it is different in China, where the processing industry is dominating, and its consuming products now dominate the world market. India adopts the elite promoting strategy, and its education system is closed linked with economic demand; China is a country of producers of popular commodities, and it needs to shift from a "world factory" to a national economy founded on knowledge and high technologies.

The subprime crisis in the United States in 2008 triggered a worldwide financial crisis, with drastic depression in American economy. With this as an opportunity, some scholars started thinking the fall of the United States and rise of emerging Large Countries. In his new work *The Post-American World* (2009), the American scholar Fareed Zakaria pondered upon the new order of rise of Large Countries. He vividly described a series of events: "Just one month in 2008, India resolutely refused the proposal from the United States about Doha Trade Talks, Russia took part of Georgia, and China sponsored the most spectacular and expensive Olympic Games in history". In the newly rising Large Countries of China, India and Brazil, the basic economic volume is massive, and they all have considerable economic activities of their own (from domestic demand). Of course, despite the economic depression and rise of other countries, the global economic shares of the United States still remain stable. It can be seen that the era of new economic order with rise of Large Countries will be one that the United States and the newly rising Large Countries China, India, Russia and Brazil coexist and fuse together.

2.2 Views of Domestic Scholars

2.2.1 Views of Large Country Economy

In China, Prof. Zhang Peigang, a famous development economist was the first paying attention to the development of Large Country economy, he stated in the book *New Development Economics* published in 1992 that, development economics should "highlight the focus of research and lay emphasis on the research of developing Large Countries". This is because since the end of WWII, except for the petroleum exporting countries that quickly became high and middle income countries with their special resources conditions, there are 11 emerging industrialized countries (regions) fairly successful in their development. They all have quite limited territory and population except Brazil. Therefore we can say that the theory and policies of development economics were successful in very few medium and small countries and regions in the third world without typical significance, while late-developing Large Countries including China and India have not yet realized steady takeoff and rapid development up to then. In Chap. 3 of the book, he "tried to make further exploration on issues of development of Large Countries, characteristics of Large Countries, difficulties in the development of Large Countries and special roads of Large Countries". At that time, Zhang Peigang did not clearly put forth the concept of Large Country economy, but he considered issues of development in Large Countries more in the economic point of view, therefore his ideas actually set the foundation for the research of Large Country economy. After that, Chen (1994) extended the research by Zhang Peigang, and conducted relatively systematic research on the development of Large Countries. Chen Wenke can be said as one of the first scholars making complete exposition on the development contradictions in late-developing Large Countries. Chen (1994) revealed ten major contradictions in the development of Large Countries according to the internal logics: population puzzle, structure puzzle, ecology puzzle, disaster puzzle, shortage puzzle, debt puzzle, market puzzle, system puzzle, puzzle of farmers as well as puzzle of traditional culture. In his view, these ten major puzzles can be classified at least into three categories: the first ones are basically economic factors; the second ones containing both economic and non-economic factors; and the third ones are comprehensive, incorporating political, economic, social, cultural and historical factors. Although the development economics represented by Zhang Peigang and Chen Wenke conducted relatively complete researches on the question of development in Large Countries, it is regretful that neither of them put forth the concept of Large Country economy, and their researches were made only on the surface of the issue, without systematic analysis of the characteristics, advantages and development strategy of Large Country economy.

In China, it was Li You and Tong Youhao who addressed the issue of Large Country economy in a fairly systematic way. Li (2000) analyzed the country scale and its relationship with the market structure, resource endowment, regional economy, economic opening, industrial policies and management system, revealing the characteristics and patterns of economic development and governmental administration under the restriction by country scale. However, his purpose of research was revealing the characteristics and patterns of economic development and governmental administration in Large Countries, instead of laying the emphasis on the advantages and scale of Large Country economy. Youhao (1999) explored on what Large Country economy is, the main characteristics of Large Country economy, the role of domestic market in Large Country economy, the relationship between Large Country economy and opening economy as well as what economic development strategy China should choose after taking the initial shape of Large Country economy, and this laid a good foundation for the research on Large Country economy. However, Tong Youhao did not go further along his fairly systematic thinking, therefore, no scientific answer was obtained to questions as what economic entities are Large Country economy and the origin of advantages of Large Country economy.

2.2.2 Characteristics and Advantages of Large Country Economy

All researchers on Large Country economy agree that Large Country economy has its own characteristics, with advantages beyond comparison by other economies, but they have not reached agreement on what are the characteristics of Large Country economy and the origin of Large Country advantages.

Youhao (1999) believed that, Large Country economy has characteristics different from small country economy: vast territory, rich resources, large population, huge domestic market, large economic scale, great influence on world economy, and complete industrial sector system. These characteristics of Large Country economy have determined its advantages, that is, a massive domestic market, the advantages of which play an extremely important role in the economic development of the country. Fengde (1988) also accepted this view, while summarizing the general characteristics of Large Country economy, he pointed out that market potential is the foundation of Large Country economy, however, the market potential of Japan is not big as compared with that of the United States and Russia, then why it has become a economic power? So the root cause lies in the market purchasing power, the domestic market of Japan is small, but its market purchasing power is high.

Zhang (2007) stated that by Large Country economy, it refers to an economy entity with national territorial area exceeding 1 million square km^2 and population exceeding 100 million. Because only with certain national territorial area, the resource endowment structure can be relatively reasonable, to form a complete domestic system for division of labor; only with a given total population size, can there be certain market purchasing power, so that the characteristics and advantages of Large Country economy can be embodied. Large Countries with these characteristics mainly include: Russia, United States, China, Brazil and India. Unbalanced development of productive force with hierarchic levels is a main characteristic of Large Country economy. On this basis, he summed up advantages of Large Country economy on main aspects: resources, division of labor, labor force, market, technological development, scale productive force advancement and resources integration. These advantages function very well for China to give full play to its advantages of Large Country economy, realize the transfer of economic growth centers and boost sustained and steady economic growth.

The Large Country Economy Research Group of Hunan University of Commerce with Ouyang (2006) as an academic leader put forth the concept "comprehensive advantages of Large Country", and also expounded in a fairly clear and systematic manner its connotation and characteristics. They held that the greatest characteristics of Large Country economy are diversity and adaptability. Diversity has led to difference in development levels, such as differences in technologies, human capital, capital accumulation and regional trade, and by adaptability, it means that the technologies, human capital, capital and trade at different levels can just suit their extent of development, therefore promoting the rapid development of regional economy. On the basis of diversity and adaptability of Large Country economy, they pointed out that the advantages of Large Country economy is comprehensive, and is a special advantage integrating the advantages in technologies, human capital, fund and trade, that is, the comprehensive advantages of Large Country (CAOLC). The theory of CAOLC is about the composing factors and occurring mechanism of CAOLC and its application in regions, industries and enterprises. Their theory about CAOLC not only combines with the views of Zhang Lijie, Tong Youhao and other people, but also covers the researches on issues of characteristics of Large Country economy, origin and functioning mechanism of CAOLC, and development strategy of Large Country economy in a complete and systematic way, therefore representing the systematic researches of Large Country economy in China.

2.2.3 Patterns and Strategies of Large Country Economy

Large Country economy has advantages and characteristics different from those of other economic entities, therefore, its development patterns and strategies also have their own characteristics, of course, such characteristics must be based on the reality and foundation of Large Country economy.

Jing (2000) studied development patterns of Large Country economy, and stated that in Large Countries all over the world, the economic growth is mainly pushed by the domestic market demand. In their active participation in international economic and trade exchanges, economic development has always been focused on domestic demand, and mainly depending on the expansion of domestic demand is a basic pattern of the development of Large Country economy, and it is also a typical demand pushing pattern. According to Zhang (2001), the main supporting force for economic growth in Large Countries is the domestic market demand, and consumption contributes most to economic growth, with investment being the main driving force of economic growth; his view is investment pushing pattern, which has advanced further on the basis of demand pushing pattern.

Youhao (2001) stated that, Large Country economy has a huge domestic market, which makes the international market less important, resulting in less opening to the outside world, therefore efforts in the development of foreign trade can more effectively promote the development of Large Country economy. The foreign trade pushing pattern of Tong Youhao has been favored by scholars, Zhang et al. (2006) analyzed the mechanism of trade and opening up as well as out-bound direct investment promoting regional economic development in the view angle of Large Country economy, and also explored on the effect and difference of trade and opening up as well as out-bound direct investment in the east, central and west regions of China, taking China as an example. Researches have shown that, opening up is conducive to knowledge accumulation and economic growth, however, custom tariff reduction produced the most obvious effect on the east region, while the preferential policies for foreign direct investment with development zones as platforms have the most obvious function in the central region. After comparing the development patterns of Large Country economy and small country economy, Tian (2001) stated that Large Country economy should make more efforts in internationalization to achieve the economic development goals.

The development strategy of Large Country economy has long been a focus of argument in the academic circle, especially on the strategy in selecting industrial sectors. Haibing (2005) pointed out that, as an economic development strategy, Large Countries should concentrate on science and technology, culture and public health, which form the foundation for development of Large Country economy, and the Large Country economy development strategy proposed by him is a massive system consisting of national defence construction strategy, economic development strategy, education development strategy and special development strategy.

Economic strategies must be finally materialized in the selection of specific industries. Qi (2006) supposed that in a country, there are only two types of industries: one is labor-intensive and the other knowledge-intensive, and he referred in brief the development pattern dominated by labor intensive industries as "China pattern", and that dominated by knowledge-intensive industries as "India pattern". Then he respectively established labor force market models for labor intensive industries and knowledge-intensive industries, to compare and research the China and India patterns. His conclusions show that, to have more people enjoy the welfare of economic development, the China pattern is better than the India pattern.

Because the China pattern enables both labor-intensive and knowledge-intensive industries to coordinate and fully develop, and can solve the problem of unemployment to the maximum extent. This research result has fairly strong reference significance, and also well explains the China pattern and India pattern, the two entirely different Large Country economy development strategies. Yang and Yang (2007) put forth the concept of "limited catch-up", and measured and calculated the limited catch-up indices of 112 countries during 1965–2000. Their cross-nation analysis showed that, in Large Country economies, countries adopting "limited catch-up" have achieved development results obviously better than those not adopting this strategy, and it also indicated that while developing comparative advantage industries, China also endeavors to develop some medium and high end industries, which is in line with the demand of "limited catch-up", therefore has achieved good results in economic development.

Xu and Ma (2007), on the basis of comparing the difference between Large Country economy and small country, further stated that, Large Country economy, with vast territory, rich resources, large population, domestic market and huge economic scale, exerts major influences on the world, therefore it is appropriate to adopt the import substituting strategy to actively expand domestic demand, without excessively depending on foreign trade.

2.2.4 *Comparative Study of Large Country Economies*

Large Countries in the world have their own particularities as well as some things in common, therefore, it is necessary to perform comparative study for Large Country economies. This study is represented by Fudan University World Economy Research Institute, the research group (2006), with the big economic countries in the world as objects, made an overall analysis of the tracks of development of Large Country economies such as the United States, Japan, Germany and Russia since the 1990s, and made clear the institutional transition in the development of Large Country economies and the background, measures and effects of structural readjustment, providing very good reference to the Large Country economy development in China. In the meantime, Cheng and Li (2006), taking the United States, Soviet Union, Brazil, India and China as objects, studied their experience and lessons in economic development, which are of more direct reference value to the economic development in China. Bai (2008) discovered by comparing the economic rising of China and India that, although the two countries selected two different roads, they now face common difficult issues in their development, and the root cause is the extra-large scale of the country. The rise of China and India calls for late-developing Large Countries to innovate development patterns, and it is a global subject of the times to explore roads for sustainable development of economy in Large Countries like China and India. Wang (2008) conducted comparative study on the economic development in the United States and Brazil, analyzed in detail the general conditions of economic development and economic development patterns of both countries, and also did detailed comparison of the development of agriculture and industries in the United States and Brazil; he explored on the different roles of the governments of both countries in forming the economic development factors, and analyzed the causes of rapid economic development in the United States and slow development in Brazil, with the conclusion that it was related to the importance attached to the role of market in the United States and negligence of the role of market in Brazil. Shen et al. (2008) conducted comparative study on the economic development in China and India, and stated that both China and India are Large Countries with large population and territory, and also developing countries with most rapid economic growth in the world today, but both of them face the test of lagged urbanization, increasing environmental pressure and expanding income distribution gap; these problems are mainly on the institutional and social levels, only these bottlenecks are broken can they get the development space in the future and improve the quality of economic and social development. On this basis, Lu (2008) made an overall analysis of the urbanization strategy, globalization strategy, industrialization strategy in the economic development in Large Countries, together with systematic carding of the political and social foundation for Large Country economy development, providing very good reference to the Large Country economy development in China.

These comparative studies are qualitative. Today with daily increasing economic globalization, are there certain links between Large Country economies, and has the economic fluctuation between Large Country economies something in common? Quite few scholars gave their empirical answers to these questions with quantitative analysis as a tool. Song and Fang (2007) analyzed the interaction and influence of economic fluctuation in China and the world, and their results obtained after a series of metering and tests showed that, the correlation of China economy and world economy experienced a course of strong to weak and the gradually becoming strong again, but the overall correlation is fairly low; the economic fluctuation in China economy lagged behind that of world economy, indicating a fairly heavy dependence of China economy on world economy; and at the 10 % significance level, the two are in Granger causality, but the world economy has greater influence on China economy. There are many linking ties between Large Country economies, and FDI is one of the most important, Zong (2007) chose this tie, and made empirical analysis on how the fluctuation of economy in world Large Countries represented by G7 influences the economic fluctuation in China via the FDI in China by using such metering methods as cointegration, VAR impulse response and variance decomposition, on the basis of the economic data of 1980–2006. The results showed that the FDI growth rate in China is correlated to the GDP growth rate in China; in China's FDI variance decomposition, components explained by economic factors of China are much more than those explained by economic factors in other Large Countries in the world, and in the GDP variance decomposition of G7 and China, components explained by economic factors of the opposite side are on the trend of rising, indicating daily deepening mutual influence in economic fluctuation between China and the other Large Countries in the world.

2.2.5 The Large Country Economy Strategy of China

China is facing problems much more severe and complicated in its economic development than those faced by the United States and other developed countries, and they must be dealt with squarely in developing the Large Country economy in China. Jinyong (2005) noticed frequent trade frictions between China and the United States and between China and Europe, with occasional intensifying, and he pointed out after surveying the important roles of other free trade areas on Large Country economy development strategy that, to go to a Large Country economy, China must adopt the form of free trade area to combine with other economic entities, so as to effectively cope with trade frictions and conflicts in the course towards a Large Country economy, and provide a favorable environment for the economic development. Gao (2007) conducted systematic researches on the problems and challenges faced by China in developing Large Country economy, and stated that the Large Country economy of China should accomplish the dual transformation from a planned economy to market economy and from an underdeveloped economy to a developed economy, this has determined that the Large Country economy of China will face not only bigger tests from overseas pressure, but also potential challenges from the uncertainty of internal development, which are mainly demonstrated in 6 aspects of industrial structure, regional difference, resources utilization, environmental protection, domestic consumption and trade structure; he further stated that the basic schema to deal with external challenges for China to advance to a Large Country economy should be transforming the development ideas and economic growth pattern, so that the economic growth in China can change from mainly depending on expansion of factor input to mainly depending on increase of efficiency.

As a developing Large Country economy, what development strategy should China select? This is a long debated topic, and also a hot question in the academic circle researches in recent years. To sum up, there are mainly four viewpoints.

1. Catch-up strategy
   This strategy originates from the theory of later-mover advantage. Lu (1999) put forth the "development motive force theory" framework based on later-mover advantage, stating that by learning-based catch-up, the development gap between China and early-mover countries will be narrowed gradually, but a certain constant gap will remain at all times; this requires us to upgrade and change the motive force of development, i.e., from the previous importation and learning to self-reliance innovation. Xibao (2002), starting from the basic theory of development economics, studied the economy catch-up theory of the west, and stated that China should take a catch-up strategy based on later-mover advantage, and proposed a complete framework on aspects of human capital, technological innovation and institutional innovation. Gaolan (2003) studied the relationship between later-mover advantage theory and economic catch-up strategy, and concluded that it is the effective use of later-mover advantage and following the general law of economic catch-up that created the "China Miracle". On the basis

of the later-mover advantage theory, many economists held that developing countries, especially China, as a late-developing Large Country, should adopt the catch-up strategy in economic development. However, quite a few scholars, such as Lin and Cai (1999), stated that catch-up strategy has some defects in itself: this strategy will result in growth in developing countries being mainly in industrial sectors, leading to an unbalanced industrial structure; the modern industries built up in catch-up strategy cannot well combine with the traditional industries and small and medium-sized enterprises, resulting in dual economic structure, and industrial sectors cannot play a linked role in driving up economic development; the industries built up in this way can hardly have self-production capacity and will be lacking in competing power; the intensive interference of economy by the government will seriously distort resource allocation, making market allocation difficult to play its role.

2. Comparative advantage strategy
   Lin and Cai (1999), Lin and Li (2003) on the basis of Ricardo's concept of "comparative advantage", stated that by adopting catch-up strategy, governments of developing countries choose to develop in priority industries and technological structures not matching with the comparative advantage determined by factor endowment, and it will result in lacking of self-surviving capacity in sectors developed in priority on a competing market. However, the governments, to enable these enterprises lacking in viability to establish and survive, must distort interest rate, exchange rate and prices, and allocate resources by administrative means to subsidize or protect these enterprises. So the market role is suppressed, rent-seeking behavior prevails, and the end outcome will be very poor performance in economic development. Therefore, the best strategy for backward countries to catch up with developed countries is the comparative advantage development strategy. After that, many scholars in China conducted empirical researches on the comparative advantage strategy. The empirical researches by Zhang and Xiaozhong (2001), and Tao (2005) verified that the revealed comparative advantage index of China has gradually matched with its endowment of labor resources. Yu and Gu (2005) expounded the transformation of factor endowment structure in China resulted from the application of comparative advantage strategy under the conditions of open economy and the transform of industrial and trade structure arising from it, analyzed the logic relationship between foreign trade, circulation of international production factors, economic growth, changes in resources factor endowment as well as economic structure changes by means of HOV model and metering model, and pointed out the guiding position of the utilization of comparative advantage and its dynamic principle in the economic growth and structural regulation of China. In addition, scholars such as Wei (2001), Hou and Jiang (2004), Zheng and Ren (2005), Guan (2007), Yang (2007) conducted in-depth researches on it, and they pointed out that in the economic development of China, the comparative advantage development strategy should be adopted, to realize leap-forward development of economy at the minimum cost.

3. Comprehensive comparative advantage strategy
   This strategy is an important viewpoint of the new classical economics, that is, every country should make full use of the comprehensive comparative advantages formed by exogenous comparative advantage, endogenous (division of labor) advantages and transaction efficiency comparative advantage, and avoid trade with products with comparative disadvantage. Yang and Zhang (2001) stated that the viewpoint of comprehensive comparative advantage has taken into account the factors neglected by the afore-said catch-up strategy and comparative advantage strategy, and therefore should be affirmed, however, they are also questioned by other economists because they took the transaction efficiency as the only prerequisite for economic growth and the constitutional framework as the only condition for economic development. Wei (2004) put forth a hypothesis that the competitiveness of a region is jointly determined by comparative advantage and competitive advantage, stating that the regional competitiveness is determined by the regional resource endowment differences as well as the enterprise competitive advantages at microscopic level, and is the outcome of comprehensive effect of them, they also classified the regional development strategy into four basic types on the basis of the magnitude of regional comparative advantage and enterprise competitive advantage, as a basis for China to select its Large Country economy strategy.
4. Strategy of comprehensive advantages of large country
   Amidst the debates over economic development strategies for developing countries, the Large Country Economy Research Group of Hunan University of Commerce put forth the strategy of comprehensive advantages of Large Country, in studying the Large Country economy, they noticed the differences, stages and adaptability in the development of Large Country economy, and systematically stated that the comprehensive advantages of Large Country originate from the diversity and adaptability of technologies, human capital and regional differences, the comprehensive advantages of Large Country has integrated the comparative advantage and competitive advantage, early-mover advantages and later-mover advantage, providing a basis for Large Countries to work out economic development strategy in a scientific way. The theory of comprehensive advantages of Large Countries shows that, the unbalanced development of regional economy requires selection of different regional economy development strategies, and this has made clear the direction for the coordinated development of regional economy, and also provided a basis for industrial selection and development.

2.3 Summary and Reviews

It can be found in summarizing the above analyses that, there is a long history in foreign countries in the research of country scale and Large Country economy: Simon Kuznets analyzed the effect of the size of country on structure of sectors,

Hollis B. Chenery studied the types of Large Countries and their scale effect, and some other scholars studied the economic characteristics of Large Countries as well as the laws of rise and fall of Large Countries. However, probably because factors involved in Large Country economy are complicated, it is difficult to set restricting conditions, and especially it is not easy to make analysis with numerical models, no scholar has made systematic research on Large Country economy, and comparatively speaking, there have been more researches on small country economy. For example, Hollis B. Chenery studied the small country pattern of economic development, Robert. J. Barrow specifically studied the inventory and output behavior of a small country economy, Gillis et al. specially studied the balance of open economy in small countries.

The economy today is becoming dominated by Large Country economy, and the daily frequent economic ties has made the research of Large Country economy a key for us to understand the world economic development, therefore, Large Country economy is gradually becoming a hot field of research in the academic circle. In particular, new progress has been made in the Large Country economy research since Goldman Sachs of the United States, in search of strategic opponents, put forth the concept of BRIC as an emerging market in the world. Since the 1990s, domestic scholars have made a lot of research on Large Country economy, and reached common view on four aspects:

First, Large Country economy has an important position in the world economic situation. They have realized the importance of Large Country economy, and that Large Country economy has become an engine of world economic development, therefore, grasping the law and characteristics of development of Large Country economy means grasping the impulse of world economic development.

Second, Large Country economy has significant characteristics different from small country economy. They have analyzed the characteristics of Large Country economy on national territorial area, population size, resources conditions, market potentials, and economic scale, including the differences and diversity of economic development factors of Large Countries, and are gradually coming to the essential characteristics of Large Country economy.

Third, Large Country economy has the obvious advantages in pushing ahead economic growth. They have come to know the existence of Large Country advantages in the reality of rise of Large Countries, studied the comparative advantage, later-mover advantage and comprehensive advantages from different points of view, and made initial analysis of the internal mechanism of Large Country advantages promoting economic development.

Fourth, Large Country economy has a development pattern different from small country economy. They have initially cognized the characteristics and unique advantages of Large Country economies, the development strategy patterns determining the Large Country economy are different from those for small country economy, and Large Country economy must choose a development strategy built on this advantage foundation.

However, in general, scholars both at home and overseas have not made sufficient systematic and in-depth research on Large Country economy, and they are not quite clear about many questions, lacking in scientific definition, therefore, the researches on Large Country economy are not systematic, only at shallow level. Specifically, obvious defects can be found in the following aspects:

First, there is no clear boundary and classification for "Large Country". Some equal "Large Country economy" to "big economic country", and some equal "Large Country" to "powerful nation", but no scientific classification criterion and evaluation system has been put forward. When studying the history of rising of Large Countries, many scholars referred some economically and politically powerful countries as "Large Countries", these countries are not big in area and population, and actually they are "powerful nations". Some scholars put forth the ideas of "big petroleum country" and "big iron and steel country", but actually they were talking about "big industrial country" or "big economic country". A small number of scholars defined "Large Country" or "late developing Large Country", but the definition seems not complete and specific.

Second, the essential characteristics of Large Country economy advantages have not been well grasped. Superficially, Large Countries have vast national territorial area, fairly large population and rich natural resources, and these are the realistic foundation for the research of economic characteristics of Large Countries. However, there are essential things at deeper levels under these phenomena, for example, the scale advantages and advantages of division of labor derived from the characteristic of "big size", the adaptability and compatibility resulted from difference and diversity, as well as independence and stability resulted from large scale. In short, difference, large scale and adaptability are critical factors in forming the Large Country economy advantages. But it is a pity that present researches have not realized the compatibility and adaptability in Large Country economy development factors, and have not well grasped the essential characteristics of Large Country economy advantages.

Third, no systematic strategic pattern has been put forth for Large Country economy development. There is obvious unbalance in the development of Large Country economy, therefore, suitable development patterns and strategies must be adopted according to the reality of difference, to effectively promote the coordinated development of regional economy. However, most of the present researches try to pursue for a unified pattern and strategy, neglecting the differences and adaptability of development strategies and patterns. Furthermore, in the research of Large Country economy development in China, scholars always start from certain theories in foreign countries, either intentionally or unintentionally, and then apply this theoretical framework to China, instead of really analyzing the characteristics and development pattern of Large Country economy on the basis of the national conditions of China.

To overcome these defects and make the researches more systematic and deeper, we need to strengthen the research on Large Country economy, mainly on three aspects: the first is to deeply and systematically define and study Large Country economy on theory; the second is to study deeply and systematically the particularity and universality of Large Country economy; and the third is to study deeply and systematically Large Country economy characteristics and strategic pattern of China. After grasping the above three points, we can further expound and test the basic theories about Large Country economy, and explore on Large Country economy development patterns that suit the national conditions of China.

Notes

1. Adam Smith is the originator of modern economics, and almost all economics theories can be traced to his *An Inquiry into the Nature and Causes of the Wealth of Nations*. For theories of Large Country economy, origin can also be found in his analysis of the relationship between scope of market and division of labor. Refer to: Adam Smith: *An Inquiry into the Nature and Causes of the Wealth of Nations*, the Commercial Press 1972 Edition, page 16.
2. [UK] Alfred Marshall: *Principles of Economics*, the Commercial Press 2005 Edition, page 291.
3. [UK] Arthur Lewis: *The Theory of Economic Growth*, the Commercial Press 2002 Edition, page 12.
4. Although Chandler analyzed the function of scale and scope on economic development from the viewpoint of industries and enterprises, it has reference value to the study of the role of economic scale of countries. Refer to: Alfred Chandler Jr.: *Scale and Scope: the Dynamics of Industrial Capitalism*, Huaxia Publishing House 2006 Edition, page 16–17.
5. [US] Kuznets: *Modern Economic Growth*, Beijing Institute of Economics Publishing House 1989 Edition, page 14, page 262–265.
6. [US] Kuznets: *Economic Growth of Nations*, the Commercial Press 1985 Edition, page 144–152.
7. [US] Hollis B. Chenery, [US] Moises Syrquin: *The Patterns of Development*, Economic Science Press 1988 Edition, page 94–126.
8. Same as [US] Hollis B. Chenery, [US] Moises Syrquin: *The Patterns of Development*, Economic Science Press 1988 Edition, page 107.
9. [US] Chenery, [US] Syrquin: *Industrialization and Growth: A Comparative Study*, Shanghai Sanlian Bookstore, Shanghai People's Publishing House1995 Edition, page 87–88.
10. Same as [US] Chenery, [US] Syrquin: *Industrialization and Growth: A Comparative Study*, Shanghai Sanlian Bookstore, Shanghai People's Publishing House1995 Edition, page 101.

Chapter 3
Concept: Comprehensive Advantages of Large Countries (CAOLC)

This chapter starts the research in this book. The author will make scientific definition and systematic summary of “Large Country”, and will also make detailed analysis of the Large Country effect of the BRIC. On this basis, the author will put forth the concept “comprehensive advantages of Large Country” (CAOLC), and expound the industrial competitiveness, enterprise competitiveness and foreign trade competitiveness based on CAOLC, as scientific interpretation of CAOLC.

3.1 Definition of Large Country

The phenomena and causes of sustained and rapid growth of Large Country economy is attracting daily increasing attention of people, however, up to date, there is no unanimous view in the academic circle on the definition of the concept of Large Country, while a reasonable definition of this concept is an important prerequisite for analyzing questions concerning Large Country economy such as the basic characteristics of Large Countries. On that account, the author will, on the basis of defining the concept of Large Country, make systematic exposition of the basic economic features of Large Countries and their evaluation indicator system, so as to lay the foundation for the research of Large Country economy and CAOLC.

3.1.1 The Concept of Large Country

In studying the economic issues of Large Countries, many scholars at home and overseas have defined the concept of Large Country based on the purposes of their studies. The definition can be based on two categories of indicators: first, a single or a number of statistic indicators such as national territorial area and population scale. Some scholars used a single total statistic indicator to define Large Country, for

Y. Ouyang, *The Development of BRIC and the Large Country Advantage*,
DOI 10.1007/978-981-10-0633-3_3

example, in analyzing the effect of country size on the domestic production structure of a country, Kuznets (1985) used population as a scale by classifying countries into big and small ones with 10 million as the demarcation line.[1] Chenery et al. (1995) referred a country with a population of over 20 million as a Large Country. Some other scholars used a number of total quantity statistic indicators to define Large Country, for example, Zhang (2007) referred a country with territorial area exceeding 1 million km^2 and population over 100 million as a Large Country. Youhao (1999) stated that a Large Country is one with vast land area, rich resources, large population, huge domestic market, a complete system of industrial sectors, fairly large gross economic scale, and considerable influence on world economy. Second, the relevant principles of economics, as compared with indicators of the first category, there are relatively less researches based on the second category of basis to define the concept of Large Country, and among them, the research by Zheng (2007) was fairly representative. She provided the definition of Large Country in economic significance based on the principle of economics: a country that can become a "price" setter in an "international market", instead of passively accepting set "prices". These researches already conducted have provided important references for us to define the concept of Large Country, but their main restrictions are: researches based on gross quantity statistic indicators provide no interpretation in conjunction with relevant economic theoretical basis, although they can clearly separate Large Countries from small ones; while researches based on principles of economics can usually only define Large Countries on certain aspects, such as Large Country of clock and watch and Large Country of petroleum, therefore it is difficult to reveal the basic characteristics of comprehensive Large Countries different from small countries. Therefore, the author will define the concept of Large Country on the basis of relevant economics theoretical foundation and incorporating the above-mentioned research results.

The theory of economic growth emphasizes the important influence of some initial conditions on economic development, and the neoinstitutional economics also states that the performance of economic development in all countries is closely linked with their respective special initial conditions formed in history. In short, the initial conditions in a country can produce important influence on its subsequent economic development level, because the initial conditions can restrict and influence the selection of economic development route in many aspects, therefore, it is an appropriate view angle for studying the differences in economic development level between countries to define the concept of Large Country according to initial conditions and analyze its influence on economic development. As some initial conditions such as institutional factor and culture factor cannot be easily measured, making it difficult to make international comparison, we made reference to the researches by Zhang (1992), Youhao (1999), Zhang (2007) and Zheng (2007) and other scholars, and screened out four initial conditions that can be easily quantified or ranked: size of the country, population scale, resource deposit and domestic market, and defined a Large Country as: a country with all conditions

Table 3.1 Territorial area, population size, domestic market scale and total resources of the five extra-Large Countries

Ranking	United states	China	India	Russia	Brazil
Territorial area (2006) (10,000 km^2)	963.2 (3rd place)	960.0 (4th place)	328.7 (7th place)	1709.8 (1st place)	851.5 (5th place)
Total population (2006) (100 m)	2.99 (3rd place)	13.12 (1st place)	11.10 (2nd place)	1.43 (7th place)	1.89 (5th place)
Scale of domestic market (2000) (100 m USD)	58,662.0 (1st place)	4743.6 (8th place)	2979.1 (12th place)	2115.9 (16th place)	1949.9 (7th place)
Potential value of mineral reserves (1990) (100 m USD)	298,392 (1st place)	165,616 (3rd place)	39,057 (8th place)	218,478 (2nd place)	16,160 (9th place)

Notes 1. Domestic market scale is expressed by household final consumption expenditure. 2. Gross resources are expressed by national mineral reserves (potential total value)
Sources The data of 2000 of household final consumption expenditure are taken from *International Statistical Yearbook* (2002); the data of 2006 of national territorial area and total population are taken from *China Statistical Yearbook 2006*; data of mineral reserves (potential total value) are from *Research of Potential Total Value of Mineral Reserves of All Countries* (*Bulletin of Geological Science and Technology*, 1999 Vol. 9), and the data of Russia are those of the Soviet Union in 1990

of vast territory, large population, huge domestic market and rich total amount of resources, able to become a price setter of some products on the international market and with sovereignty interests in the scope of the world. This definition has the feature that the selected conditions can reflect the most obvious differences of Large Countries from other ordinary countries, and are the biggest common characteristics of Large Countries, therefore, it enables us to define Large Countries and analyze their basic characteristics on the basis of these conditions more easily and with better interpretation.

According to the above definition of Large Country, we refer China, Russia, India, Brazil and United States as five extra-Large Countries, and their indicators and the ranking in the world are as shown in Table 3.1.

3.1.2 Comprehensive Evaluation Index System for Large Countries

Although the concept of Large Country is defined on the basis of national territorial area, population size, domestic market and gross resources, and United States, China, Russia, India and Brazil are classified as extra-Large Countries according to this definition, it is still a rough classification, because it is difficult to completely and accurately reflect the conditions of a country only with one indicator of each of these four conditions. In view of this, we tried to build a more scientific and complete comprehensive evaluation index system for Large Countries.

1. Building of a comprehensive evaluation index system for Large Countries
   To build this comprehensive evaluation index system, we should not only follow the laws and internal requirements of regional economy development and consider such factors as data availability, categorizing and calculation, but also apply the principle of being scientific, comprehensive, hierarchic, representative, operable and comparable.
   According to the above-mentioned basic principles and requirements, we built the comprehensive evaluation index system for Large Countries in conjunction with the definition of Large Country in this book and by taking into account the realistic conditions in some developing countries (Table 3.2). This comprehensive evaluation index system consists of 3 levels: the target level, criterion level and index level. The indices of a higher level are the summary of those at a lower level, and those at a lower are the detailed manifestation of those at a higher level.
   The basic indices of population reflect the promoting role of the factor of population in the economic development in a country, and are measured mainly with three indices: total population, labor quality and labor quantity; the basic indices of nature reflect the promoting role of resource endowment in the economic development in a country, and are measured mainly on two aspects of territorial area and resource reserves; the basic indices of economy reflect the promoting role of basic economic conditions in the economic development in a country, mainly covering market scale and industrial sectors, and the completeness of industrial sectors refers to the proportion of industrial sectors in a country in all industrial sectors at a given time point (according to international standard classification), and is used to evaluate the completeness of the industrial system of a country.
   In the actual application of the comprehensive evaluation index system in Table 3.2, we can first classify Large Countries and small countries according to the four basic indices of population size (D1), land area (D5), cultivated land area (D10), domestic production product (D11), then combine these basic indices with other indices, and calculate the comprehensive evaluation indices of Large Countries by determining their weight, to judge the comprehensive development potential of Large Countries.
2. Standardization of basic indices for classification of Large Countries and small countries
   To determine whether a country is a big one according to the basic indices, it is necessary to standardize the values of basic indices. We determine the relevant standards on the basis of the statistical quintile method.

Table 3.2 Comprehensive evaluation index system for Large Countries

Target level	Criterion level	Index level	Indices	Calculation method	Judging criterion
Definition of large country A	Basic index of population B1	Total population C1	Population size D1		50 million people
		Quality of labor force C2	Higher research enrollment rate D2 Per capita education years D3		
		Quantity of labor C3	Economically active population D4	Population aged at 16–64 years	
	Basic index of nature B2	Territorial area C4	Land area D5 Sea area D6		1 million km^2
		Resource reserves C5	Water resource D7 Mineral resources D8 Energy reserves D9 Cultivated land area D10		11 million ha
	Basic index of economy B3	Market scale C6	Gross domestic product D11 Economic growth rate D12 Household final consumption expenditure D13		43 million USD
		Industrial sectors C7	Completeness of industrial sectors D14		

(1) Classification standard for population size. At present, there are 180–200 countries (or regions) in the world, with a total population of 6 billion, so each country should have a population of 30 million by average. Therefore, by the quintile method, a Large Country should have a population of over 50 million, a small country a population of below 10 million, and a medium sized country a population of 20–40 million.
(2) Classification standard of territorial area. With the total land area in the whole world of 130 million km^2 (excluding the Antarctic), each country should have a land area of 600,000 km^2, therefore by the quintile method, a Large Country should have an area of over 1 million km^2 and a small country an area of below 200,000 km^2.

(3) Classification standard of cultivated land area. According to the database of the World Bank, in 2005, the cultivated land area in the whole world was 1421.17 million ha, or 7.10 million ha per country, therefore by the quintile method, a Large Country should have cultivated land area of over 11 million ha, and a small country should have less than 2 million ha.

(4) Classification standard of GDP. In 2007, the world total GDP was 5434.7 million USD, or 26.60 million USD per country by average, therefore by the quintile method, the GDP of a Large Country should be over 43 million USD, and that of a small country below 8 million USD (本书此节及相关部分的GDP美元值显然是不对的,请编辑核对—译注).

3. Determination of weight

Reasonable determination of the weight of all indices is critical to the calculation of comprehensive evaluation indices for Large Countries. To effectively avoid one-sidedness in weighting, we used the method of combining analytic hierarchy process (AHP) with expert scoring to determine the weight of all indices in the comprehensive evaluation index system for Large Countries. AHP is a method combining qualitative and quantitative methods, and is mainly used to deal with questions with unclear mechanism and difficult to precisely measure and quantify in complicated systems. Its basic principle is to decompose a complicated issue into a number of levels, judge and give scores to them level by level by comparing the importance degree structure judgment matrix in pairs, i.e., to determine the contributions of lower level factors to the upper level ones using the characteristic vectors of the judgment matrix, so as to obtain the ranking of importance of basic level factors to the overall target. In general, the AHP can put together the judgments of different experts and decision-makers, and process them by quantifying and centralizing, therefore it can fairly fully reflect the actual status of issues. The weight of indices in criterion level and the composite indices in the index level is determined using the AHP method, and specific indices in the index level are weighed by expert scoring. Furthermore, to reduce the effect from subjective factors, objective weighting methods such as entropy evaluation method can also be used to estimate the weight of relevant indices.

In this section, the concept of Large Country is defined on the basis of the important effect of initial conditions on the economic development in a country in the economic growth theory and neoinstitutional economics, and the basic characteristics of Large Countries are analyzed on this basis, to build a comprehensive evaluation index system of Large Countries. This has laid a fair foundation for the relevant researches in this area, however, further researches will be required on how to further systematize and quantify the definition of Large Country and its evaluation indices as proposed by the author, what indices can be used as better substituting ones for the total quantity statistic indices we are concerned about, and how the weight of indices can be determined in a more scientific way so as to obtain more objective comprehensive evaluation indices for Large Countries.

3.2 Large Country Effect on Economic Growth

In view that China, Russia, India, Brazil and United States rank in the front of the world in terms of territorial area, population, resources and domestic market, and their outstanding performance on economic operation, the author will analyze the basic characteristics of Large Countries and the Large Country effect of economic growth mainly on the conditions of these countries.

3.2.1 Basic Characteristics of Large Countries

We have summarized the following basic characteristics of Large Countries on the basis of the important effect of conditions of vast territory, large population, huge domestic market and rich gross resources on the development of Large Country economy:

First, the relative completeness of national economy system. Vast territorial land, rich resources, large population, and huge domestic market provide the necessary material resources condition, human resources condition and market condition for Large Countries to establish a relatively independent national economy system with a complete range of sectors, and also ensure their normal operation. Specifically, Large Countries have vast territories, with large space for their economic activities and industrial layout; Large Countries have rich resources (including natural resources and human resources), so that the cost of some of their production factors is relatively low; Large Countries have huge domestic market, conducive to the evolution of division of labor and specialized production. Therefore, compared with other countries, Large Countries can usually give better play to their own comparative advantage and scale economy advantages, so as to build up relatively complete and independent systems of national economy. For example, after the WWII, Brazil achieved substantial development of economy, and built up a fairly complete industrial system by implementing the "import substituting" strategy. After the war, India also achieved remarkable economic development, it gradually built up an independent and complete industrial system with heavy industry as the core, and reached fairly high standard in sectors of iron and steel, metallurgy, heavy machinery, automobile, coal, ship-building, power generation, atomic energy, chemical industry, military industry and communication, and some have even ranked in the front in the world. Russia has a profound industrial foundation, with a complete range of sectors. After the founding of New China in 1949, great changes took place in the social and economic conditions of China after over 4 decades of development, and China has built up an independent industrial system and national economy system with fairly complete sectors. But small countries normally are lacking in natural resources of various categories required in industrialization, making it difficult for them to form complete industrial chains and distribution, for example, Singapore and Korea only develop comparative advantage industries or are specialize in making products with their own comparative advantage.

Second, the unbalance in economic development. Large Countries have vast territories, and there are fairly big differences in geographic location, resources and population distribution in various areas, therefore big differences usually exist in economic development between areas, with more obvious feature of unbalance than other countries. For example, although the economic development in different areas in the United States has become fairly balanced, it was a country with quite unbalanced regional economic development in history. In 1994, Alaska had the highest per capita GDP in the United States, while Mississippi was the lowest in per capita GDP, with the per capita GDP of the former being 3.2 times that of the latter. Also, according to the official statistic data of India, during 1980–2000, the absolute difference in per capita GDP in 16 regions of India kept on expanding, while the relative gap was narrowed in the 1980s but expanded in the 1990s, the per capita GDP in rich areas being 5.6 times that of poor ones.[2] The development in Brazil is quite unbalanced between regions, its east coastal area is equal to Southern Europe in development level, but its central and western areas are still in a backward condition, and the productive force is low, dominated by agriculture.[3] There are also obvious differences between regions in the economic development in China, the ratio of actual per capita GDP of the eastern region to that of central and western regions increased respectively from 1.29 and 1.58 in 1978 to 1.99 and 2.63 in 2005; and in actual per capita GDP, the gap between the lowest Guizhou from the highest Shanghai also expanded from 5 times in 1978 to 11 times in 2005. In terms of growth rate, all 7 provinces with the most rapid actual per capita GDP growth rate during 1978–2005 are all in the eastern region, with the highest being Jiangsu at 11.13 %, and the lowest being Qinghai at only 5.7 %, and it is also only 6.19 % with Heilongjiang Province.[4]

Third, the bidirectional influence with world economy. There is no doubt that changes in world economic situation will impact the economy in various countries, but small countries usually can only passively withstand the impact from fluctuation in world economy, while the economic development in Large Countries can in turn produce major effect on world economy, that is, there is a bidirectional influence between Large Country economy and world economy. Large Country economy has large gross quantity, with complete lines of products, high supplies and huge domestic market, therefore it can produce major influence on the prices of some products and even raw materials on international market, in the meantime, given the political and military positions of Large Countries in the world, Large Countries can also produce big influence on world economy. The United States, with its total scale of economy ranking at the first place in the world, is the locomotive driving the global economic development, and its economic movement can produce far-reaching influence on world economy. For example, after the slump of US economy in the second half of 2000, most of trade partners of the United States, such as the EU, Japan, Canada and Mexico, suffered economic slump or recession, and the world economy also experienced the first synchronous downturn in the past 10 years.[5] The China's economy is gradually growing into one of the important driving forces in world economic growth, according to statistics of the World Bank, during 2003–2005, the average contribution rate of economic growth of China to the world GDP growth was as high as 13.8 %, ranking at the second place in the world only next to that of the United States at 29.8 %. The proportion of India's GDP in the world total GDP increased from 1.12 % in 1993–2.09 % in 2007, and its rising indicates gradual growing influence of the India's economy on the world economy.

Fourth, the relative stability of economic development. Large Countries, with large territorial area, rich resources, large population, relatively complete and independent national economy system and huge domestic market, are able to establish internal circulating systems for economic development, with fairly strong ability of self regulation and capacity to resist various risks and impact from world economy fluctuation, to ensure steady development of economy. Stable economic development can in turn enable Large Countries to accumulate huge amount of capital, to further enhance their capacity to resist major internal and external impacts (such as natural disaster and financial crisis). In 1998, flood disasters hit a total of 29 provinces, autonomous regions and municipalities of China to different extent, causing direct economic losses of 255.1 billion yuan. The central and local governments quickly allocated over 15 billion yuan for disaster relief and re-construction in the disaster-affected areas. Therefore, even if the impact of Asian financial crisis remained at the time, and it was followed by the SARS crisis in 2003, the China's economy still maintained rapid growth. However, in late Aug. 1998, a flood almost destroyed the whole economy of Bangladesh, at that time, the President had to ask for help from the international society. Furthermore, after the outbreak of international financial crisis in 2008, the Chinese Government made the decision to invest 4 trillion yuan to support the development of major industries and expand domestic demand, to maintain the annual economic growth rate above 8 %. The United States also launched a bail-out package of 700 billion USD, to push forward the development of economy. By comparison, Iceland, even with its per capita GDP ranking at the front place in the world, was facing the situation of national bankruptcy. In general, Large Country economy has strong resistance against risks, with the characteristic of relatively stable development.[6]

3.2.2 *Large Country Effect on Economic Growth in BRIC*

Large Country economy has advantages as well as disadvantages. This section mainly studies the advantages of Large Country economy, especially, the positive significance of development of Large Country economy is analyzed in detail on the basis of the basic national conditions and development reality in BRIC. Large Country economy has its own characteristics and laws, and has formed a Large Country effect that can boost economic growth; specifically, they are demonstrated by the advantages favoring economic growth formed by such factors as market potentials, gross resources, economic scale as well as product diversity, differences of regions and economic completeness.

1. Big market potentials: domestic demand drives economic growth
   "BRIC" are Large Countries on the emerging market, with very high market potentials, and their domestic demand can greatly pull up the economic growth; the data in *International Statistical Yearbook* (2002) show that, the household final consumption expenditure in 2000 was respectively 494.99 billion USD, 474.36 billion USD, 297.91 billion USD and 211.59 billion USD in Brazil, China, India and Russia, ranking at the front places in the world, and it still keeps on rising in recent years. This market potential is a basic factor of Large Country economy, and the massive size of market has become an important advantage in economic development.
2. Large gross resources: a domestic factor that pushes forward economic growth
   Natural resources is a basic factor for economic growth, BRIC have very large gross resources, which has provided favorable conditions for rapid sustained growth, and vast land and rich resources are basic characteristics of Large Countries. These countries all have large inventory of natural capital, according to data provided by geological sector, the total value of mineral reserves in India, China and Brazil is respectively 3905.7 billion USD, 16,561.6 billion USD and 1616 billion USD, much higher than that of some developed countries not in big scale, such as the 290 billion US dollars of the UK and Japan and 340 billion US dollars of Germany. Large gross resources favor the formation of scale economy and fostering of pillar industries. According to the principle of comparative advantage, an economy is competitive only when it selects industries and products based on the comparative advantages of factors, so, Large Countries can have large scale of industries with large gross resources. Just for this reason, China and India have become major producers of textiles in the world with their abundant raw materials for the textile industry, and Russia has become a major energy country in the world with its rich energy resources.
3. Large scale of economy: deepening the division of labor in domestic industries
   The large market size and gross resources in BRIC have determined the large economic aggregate or scale in these countries. The main effect of economies of scale as stated by Krugman lies in: saving cost and deepening division of labor. The economic scale is subjected to restriction by production factors and market size, Large Countries have large scale of production factors scale, as well as large domestic market scale, which can lead to division of labor and specialized production, and also resources agglomeration and industries agglomeration, to push forward technological segmentation and progress, helping the accumulation of social capital and improvement of environment for economic development. Just as the *2009 World Development Report* published by the World Bank pointed out: developing countries are entering a new realm of agglomeration economy. For example, in Guangdong and Zhejiang of China, some typical phenomena of agglomeration economy have been seen, and agglomeration of software industry has started in India.[7]

4. Product diversity: enhancing the comparative advantage in foreign trade
   The BRIC countries are in the process of development and transformation, the "dual structure" or "multi-structure" of economy, the regional differences and industrial differences as well as the resulted product diversity can be converted to advantages in foreign trade under certain conditions. Large Countries have vast areas and fairly complete industrial systems, and they usually have diversified structures of export commodities. At present, the main export commodities are foodstuff, textile, chemical and mechanical and electrical products from China, software products and textiles from India, agricultural and mineral products and aerospace and aviation products from Brazil, and mineral and energy products and aerospace and aviation products from Russia. In recent years, the foreign trade structure is being upgraded, but the characteristic of diversity has remained. From a positive point of view, this structure has its benefit, as it enables proper use of the natural resources, human resources and economic and technological resources in Large Countries. For Large Countries, a diversified structure of export commodities is also in line with the comparative advantage, a Large Country has very big economic aggregate, and its one region or industry can exceed the total of a whole country or all industrial sectors of a small country, therefore, even diversified structure of export commodities can also form economies of scale and scale effect, and gain competitiveness on international market.
5. Regional differences: promoting the rapid growth of regional economy
   BRIC all have vast land areas, and there are differences in the natural resources, human resources, capital accumulation and development level in different regions. Such differences can promote rapid and coordinated development of economy. On one hand, unbalance is a force to push economic development, so that different regions can develop in competition, accelerating the development rate. In the eastern, central and western regions of China, local governments attach special importance to the development of regional economy, and phenomena of rapid growth such as "gradient transfer" and "limited catch-up" can be seen. On the other hand, the differences between regions can be converted to complementarity, to become advantages promoting the coordinated development of regional economy. In these Large Countries, different regions select economic development strategies on the basis of comparative advantage, those rich in labor resource develop in priority labor intensive industries, those with abundant capital develop in priority capital-intensive industries, and those with developed technologies develop in priority technology-intensive industries, so that industries in all regions will have relative competitiveness. Furthermore, a situation of factor complementation, industry complementation and coordinated development of economy can be realized in the whole country.
6. Completeness of economy: maintaining the steady development of national economy
   BRIC countries have complete and independent national economy systems and huge domestic markets, therefore they can establish internal circulating systems

for economic development, and build up fairly strong capacity of self regulation and to resist external economic impact, ensuring steady operation of national economy. Large Country economy has the characteristics of vast territory and rich resources, with fairly abundant resource factors to develop economy and large space for industrial distribution, with these advantages, complete industrial chains and industrial distribution can be formed and relatively independent national economy systems can be established. For example, Russia has profound industrial foundation and complete industrial sectors, and established a complete national economy system early; after the WWII, Brazil and India gradually established fairly complete industry systems by implementing the "import substituting" strategy; after the founding of New China, China established an industrial system with fairly complete sectors and independent and complete national economy system by following the policy of "independence and self-reliance". Such economic advantages can well resist impact from fluctuation in world economy, for example, in the international financial crisis triggered by the subprime crisis in the United States, all BRIC countries demonstrated good resistance against risks, and maintained steady economic development by expanding their domestic demands.

3.3 Interpretation of Comprehensive Advantages of Large Countries

The so-called CAOLC is a special advantage determined by the characteristics of Large Country and of diversity. It is a comprehensive advantage formed by integrating all favorable resources of Large Countries, merging all advantages as one. There are many factors related to CAOLC, including factor endowment, market capacity, economic scale as well as important factors of adaptability of technological level and human capita, complementarity of regional economy and completeness of economic sectors. CAOLC is originated not only from individual factors, more important, it is the integrated effect of many factors, for example, the scale characteristics of factor endowment can help form the advantage of economies of scale, the diversified characteristics of technological and human capital can form the advantage of adaptability, the regional differences can be converted into complementarity advantage, and the domestic and overseas circulating systems can ensure the advantage of stability. With the concept "comprehensive advantages of Large Country", we can go beyond the thinking frame of comparative advantage and competitive advantage, thereby examining the economic development advantages of Large Countries in broader areas from a systematic perspective.

3.3.1 *Basis for "Comprehensive Advantages of Large Country"*

The author put forth the concept "comprehensive advantages of Large Country" mainly on the basis of the economic development situations of China, India, Russia, Brazil and the United States, and firstly based on the national conditions of China. Therefore, our research will be extended to other economic Large Countries with the research of China as a basis. The economic development in late developing Large Countries has the "transformation" and "diversified" characteristics: in a longitudinal view, it is in the transformation from a developing country to a developed country, some factors have been transformed and others have not, in this period of "transformation", the traditional economy and modern economy coexist, and extensive economy and intensive economy coexist; in a transversal view, it is in a status of unbalanced development of all regions, sectors and industries, this "diversified" status is one with multiple regional structures, economic structures and technological structures coexisting, and the realistic "transformation" and "diversified" characteristics are the logic start point to form the "comprehensive advantages of Large Country".

In the case of China, we are and will be in the primary stage of socialism over a long time, and our moderate prosperity today is still not complete and at low level, with quite unbalanced development. The characteristics of primary stage of socialism are identical to those of developing countries. Vast territory, rich resources and large population are the basic characteristics of Large Countries; quite unbalanced development is a basic characteristic of developing countries. The combination of the two constitutes the basic characteristics of late-developing Large Countries. China is a Large Country, with unbalanced resources distribution and economic development, with the advantages of developed countries on some aspects and the advantages of developing countries on other aspects, specifically, the multi-structures of China's economy are demonstrated on the following aspects:

(1) The regional multi-structure. There are big differences between regions in China, they have different natural resources, human resources, technological conditions and social and economic condition, with respectively features, and some of these features are their comparative advantages. China can be roughly divided into three parts: the eastern, central and western regions, with different conditions in different parts; in each part, there are cities and rural areas, also with different conditions; and different cities also have differences and their respective advantages and disadvantages. In this way, the regional multi-structures are formed, for example, the division of three major economic belts reflects the geographic environment for the economic development in China and the macroscopic regional differences in economic strength, economic structure and economic and technological level.

(2) The multi-structure of economy. Large Countries have big regional differences, the different regional resources and conditions determine the different

extent of economic development, resulting in the multi-structure of economy. The economist Myrdal put forth the concept of "geographic dual economic" and "dual-space structure" theory. He stated that geographic dual economy is one of the basic characteristics in the regional economic development in developing countries, that is, a dual structure with economically developed and underdeveloped areas coexisting, because of the difference in the economic development of different areas. Because production factors can flow freely, if the conditions in a given area are special, and the economic growth rate is higher than that in other areas, unbalanced economic development will occur in the country, which will lead to gaps in income level and profit, forming a regional dual economic structure, in developing countries, modern industries are always first concentrated in a small number of areas, and the remaining spaces become underdeveloped edges of these areas, showing a "dual structure" or "core—edge structure", that is, a space system formed by advanced and developed core areas and backward and underdeveloped edge areas. Growth of regional economy is the outcome of interaction of many factors, mainly including natural condition, labor force, capital, scientific and technological progress and resources allocation, when these conditions are different or the ways of allocating resources are different, it will lead to different extent of economic development, finally forming the ladders in regional economic development, then the gradient transfer in regional economic development will occur, and this is the ladder structure and space transfer of productive force.

(3) The multi-structure of technology. The gradient of regional economy is closely linked with gradient of technologies, the industries in high gradient regions mainly consist of sectors in innovating phase or in booming phase, while those in low gradient regions mainly consist of sectors in the late mature phase or in declining phase. High gradient regions take the lead to apply advanced science and technologies, to form new industries, products and technologies, but low gradient regions can be late in all these aspects, thereby resulting in the multi-structure of technologies in different regions. Moreover, even if within a region, multi-structure of technologies can also occur. The "dual-economy theory" put forth by the economist Lewis is actually also a "dual-technology theory", he believed that the economic structure in developing countries can be divided into two sectors, that is, modern economy and traditional economy, or modern industrial sectors and traditional ones: the former is based on advanced science and technologies, with the characteristics of technology-intensive pattern; and the latter is based on traditional manual technology, with the characteristics of labor-intensive pattern. In the industrialization process of China, dual-technology structure has also formed within the industrial systems, a phenomenon of coexistence of modern big industries and backward small industries with obvious differences in technical and equipment level and in labor productivity within the industrial system. As enterprise scale is in certain proportion to the technical and equipment level in enterprises, therefore, enterprises can be classified into high technical and equipment level

represented by big enterprises and low technical and equipment level represented by small enterprises. Of course there are exceptions, some high and new tech enterprises have high level technologies and equipment despite their small scale. As a whole, rural industry is mainly characterized by low technological level, and its increase tends to increase the low-technology enterprises in the industrial technological structure in the country; in the meantime, with the rapid development of high and new technologies, the technical gradient within the industrial system tends to enlarge, and the dual structure of technologies become more apparent.

The diversification of regions, economy and technologies can result in comprehensive advantages in the viewpoint of integrating resources, and it can further lead to diversification of development advantages, diversification of motivations, diversification of main players and diversification of industries. Comprehensive economic gains can be obtained by implementing this "diversification" strategy. For example, the diversification based on regional economic level and factor endowment structure can give play to the traditional comparative advantage in backward areas and innovation advantages in developed areas through effective division of labor between regions; the diversification of economic structure can promote the display of labor resource endowment advantages through the flow and rational allocation of factors, and also promote the dynamic transplanting of low factor cost with high quality factors, to get advantages in competition; the diversification of technology can couple the diversified adaptive technologies with human capital, material capital, industrial structure in the region as well as regional economy development level to promote the coordinated development of regional economy. This diversified characteristics of technologies adapt to the status quo of different areas, different industries and different enterprises in late developing Large Countries, can well satisfy the technical foundation of different areas, different industries and different enterprises, and properly promote the coordinated development of economy. The comprehensive economic efficiency resulted from the "diversified" characteristics of regions, economy and technologies in Large Countries is an important manifestation of the comprehensive advantages of Large Country.

In considering the economic development strategy of China and the international competitiveness of China's economy, two important questions made me think over them deeply.

The first question: what is the theoretical basis for setting the strategy of economic development in a late-developing Large Country in transition period such as China? The Party put forth the "theory of primary stage of socialism", which means that China has established the basic system of socialism, however, because of our characteristics of large population, weak foundation and low economic development level, we must endeavor to develop social productive force and speed up the pace of economic construction. Furthermore, various theories have been put forth by economists: Arthur Lewis put forth the "dual structure theory" for developing countries and regions, which classifies the economy in developing countries and regions into two sectors of modern industry and traditional agriculture, and states that the development of the whole economy depends on the expansion of the modern industry sector, yet the expansion of modern industry sector in turn needs the rich and cheap labor force supplied from the agriculture sector; Heckscher–Ohlin put forth the "resource factor endowment theory", advocating that developing countries concentrate in the production and export of labor-intensive products, while developed countries concentrate in the production and export of capital-intensive products. Michael Porter put forth the "country competitive advantage theory", stating that a country or region should be engaged in fostering industries with high level competitive advantage of its own, and also advocating that developing countries should get rid of the vicious cycle of solely relying upon cheap labor and natural resources. There are also "gradient transfer theory" and "reverse gradient transfer theory": the former advocates supporting the rapid development of high gradient regions with ready conditions, and then advancing gradually to level II and level III gradient regions; while the latter believes that the sequence of advancement can only be determined by the needs and possibility of economic development, the economic development in low gradient regions can also realize leap-forward development and advance reversely to regions of higher gradient by importing technologies and enhancing their self-reliance innovation capacity. For these theories, we may choose some contents and principles that can be merged from them, and integrate them into an economic theory that can fully and rationally interpret and guide the economic development in China.

The second question: China is only a developing country, but why some developed countries predict that China will become a powerful economic country in the world? According to the research report of the International Institute for Management Development (IMD) in Lausanne of Switzerland and data in the *World Competitiveness Yearbook*, China's ranking in terms of comprehensive international competitiveness in the world was between the 24th and 34th places in 1999 and 2000, in a trend of "rising with shocks". Its overall level is at the front of the transitional countries, basically maintaining among the developing countries ranking in the front, even exceeding some industrialized countries such as Italy and Greece, and certainly far ahead of some developing small countries. Then, what kind of advantages allow China to possess such competence? Obviously, this competitiveness cannot be achieved solely with labor resources advantage as a developing country, instead, it should be the "Large Country" advantage and "transformation" advantage: China is a Large Country with unbalanced economic and technological development, some regions and sectors have the advantages of low-cost labor resources and application technologies, while others possess the advantages of high-tech and high capital-intensive industries; China is experiencing a transition economy, in which some regions and sectors have some advantages of developing countries, while others have the advantages of developed countries, and these different advantages can all become active factors to enhance the international competitiveness of Chinese economy. Therefore, we should integrate these advantages, merge them and inject them into the source or driving system for the economic development of China, and also to find objective evidences for us to analyze and predict the development trend of China economy.

With the sustained rapid growth of China's economy, and the continual increase of its comprehensive economic strength, how to select the economic development strategy of China and how to increase the competitiveness of economy have become hot topics of research. Different scholars have applied theories of comparative advantage, competitive advantage, later-mover advantage, industrial competitiveness theory and industries agglomeration, to research questions on economic development strategy and economic competitiveness from different angles, and obtained different conclusions. For the question what economic development strategy an underdeveloped country should select, there has been the debate between catch-up strategy theory and comparative advantage strategy theory in China in recent years. In view of the differences and controversy between different schools on the origin of economic competitiveness and development strategies, Wang (2007) put forth the concept of comprehensive comparative advantage theory, and Ji Yun (2003) put forth the view of composite comparative advantage. Shanyong and Yan (2005) proposed and justified the hypothesis "comprehensive advantage development strategy" beyond this controversy, i.e., in the course of economic development, implementing the catch-up strategy can likely go into a "catch-up dilemma", but implementing the comparative advantage strategy can easily fall into the "comparative advantage trap", therefore, only by implementing the "comprehensive advantage development strategy" with overall consideration of the comparative advantage, division of labor interest and transaction efficiency, can we ensure harmonious and healthy development of economy.

After reading some works on economics and literature commenting on the economic development in China, and thinking, analysis and summary in a deep-going way, the author put forth the basis for setting out the economic development strategy of China and predicting the international competitiveness of China's economy, that is, the "comprehensive advantages of Large Country", it refers to a special advantage determined by the "Large Country" characteristics and "transformation" characteristics of China, and is a comprehensive advantage obtained by integrating various favorable resources of a late-developing Large Country in the transformation period and merging the developing country advantages and developed country advantages into one. Specifically, China, as a late developing Large Country, has got the "comprehensive advantages of Large Countries" with distinctive features on aspects of natural resources, human capital, financial capital, technological progress, market demand and foreign trade:

Natural resources: rich + differences between regions
Human capital: cheap + fairly high quality
Financial capital: total amount + high concentration
Technological progress: simulation + self-dependent innovation
Market demand: worldwide + mainly domestic
Foreign trade: diversity + dominating products.

This shows that China not only has rich natural resources, suitable to the operation of the domestic cyclic system, and the distribution of resources is different in regions, conducive to the division of labor and cooperation of these regions; there is relative cheap human capital that can enable enterprises to cut cost, yet the human capital has an education level higher than that in some other developing countries, conducive to upgrading the technological level and management level; the total amount of financial capital is big, which can meet the capital demand in domestic development, and also enables applying the capital in a relatively concentrated way in given regions or industries, to realize large scale operation and first-mover advantages in these regions or industries; the technological progress has attached importance not only to imitation, to meet the general technical demands of the economy and society, but also to the self-dependent innovation of critical technologies, to enhance the core competitiveness of the country; the market demand potential is high, not only with huge domestic market demand, but also enabling full utilization of international market, to form an international and domestic dual-drive pattern dominated by domestic market; the foreign trade not only has a diversified export product structure, able to adapt to the consuming demands of different countries, different areas and different strata, but also has its own dominating industries and dominating products, able to build up scale advantages and competitive advantage. These special advantages formed by the characteristics of late developing Large Countries can be summarized as the "comprehensive advantages of Large Countries".

3.3.2 Industrial Competitiveness Based on Comprehensive Advantages of Large Countries

By the so-called "industrial competitiveness", it refers to the competitiveness of industries and all industrial fields of a country in the world. Industrial competitiveness is selective, because no country can gain advantage in all industrial fields. Even within a given industry, it is not possible to make all commodities, labor and technologies the most powerful in it. Therefore, a wise choice is giving play to the own advantages, so that some industries, or some links and elements in a given industry are competitive internationally. In the selection of industries, China should organize scientific and rational industrial combination with the CAOLC as the basis.

The selection of industries based on CAOLC should be established on the characteristics of regional, economic, technological multi-structures. First, China is a Large Country, and should establish a fairly complete and well balanced industrial structure in development, and the CAOLC offers the possibility to establish such a complete industrial structure, in small countries, either developing or developed, it is neither possible nor necessary to establish a complete industrial structure system, because that will be a choice lacking in objective condition and without efficiency, and it will go against the principle of optimizing resource allocation according to the mechanism of market economy in a broader range. Second, China is a developing countries, its economy and technologies are not well developed, and traditional industries and labor intensive industries still hold important positions, so the CAOLC has made it possible to bring into play the advantages of traditional industries and labor intensive industries. In any country, industries can be selected according to its actual conditions. It is not possible for us to separate ourselves from the actual situation of economic development in China, discard the traditional advantageous industries and labor intensive industries, and develop high and new technology industries and capital-intensive industries in all regions. We can only determine a comparatively reasonable "quantity", and gradually optimize our industrial structure through long-term efforts. Third, China is a developing country in transformation, and high and new technology industries have been developed fairly well in some regions; this is the hope for the rise of China, and the CAOLC has made it possible to energetically develop high and new technology industries. To rank our country among the economic powers in the world, we cannot completely follow the "industrial gradient transfer" theory, to accept the transfer of marginal industries from developed countries, because in that way, we can only pick up the pace of developed countries, but will never become a developed country forever. Only by integrating the advantageous resources of our own and making efforts to foster high-tech industries with competitive advantage, can we implement the "leap-forward" development strategy, and realize economic modernization as early as possible.

The general guideline set by the Chinese Government to upgrade the technological level of industries with self-reliance innovation, speed up the development of advanced manufacturing industry and service industry as well as strengthening the construction of basic industries and infrastructures has not only reflected the general law of international economic development and basic requirements for enhancing industrial competitiveness, but also comply with the realistic situations of China and basic industrial development orientation in the future, and is therefore scientific and reasonable. However, this pattern and idea should be concretized, to better guide the industrial distribution and practice in construction.

In the first place, we should be able to identify advantageous industries in specific regions in the market competition pattern. Under the market economy conditions, the advantageous industries in a specific region not only depend on its resource conditions, but also are the outcome of selection and competition of a large number of industrial and commercial entities. Therefore, the advantageous industries in a specific region are the industrial sectors with comparatively high share on the broad domestic and international markets. We can use the location quotient approach, and process the data of added values provided by existing statistics to obtain the "equivalent share" of all industries in the region, to reflect the market competing power of each regional industry. The calculation formula of this indicator is:

$$(1)\ LQ_{ij} = \frac{L_{ij}/\sum_j L_{ij}}{\sum_i L_{ij}/\sum_i \sum_j L_{ij}};\ (2)\ LQ_{ij} = \frac{L_{ij}/\sum_i L_{ij}}{\sum_j L_{ij}/\sum_i \sum_j L_{ij}}$$

where: i stands for the ith region, j for the jth industry; Lij the output value of the jth industry in the ith region; and LQij the location quotient of the jth industry in the ith region.

The location quotient in the first form stands for the ratio of the share of the jth industry in the ith region in the total output of the region to the share of the jth industry in the whole country in the whole national economy output. The location quotient in the second form stands for the ratio of the proportion of the jth industry in the ith region in the same industry in the whole country to the proportion of the economic aggregate of the ith region in the economic aggregate of the whole country. Both location quotient formulas can help us measure and identify advantageous industries in regions.

Second the advantageous regions of all industries and their ranking should be analyzed systematically. Research results of relevant institutions indicate that, the ranking by regions from advantage to disadvantage is the west, central and east regions in agriculture, the east, central and west in industry, and the east, west and central in service industry. Therefore, industries should be selected by different regions in the sequence of service, industry and agriculture in the east, agriculture, industry and service in the central region, and agriculture, service and industry in the west. It was found in a survey of the advantage distribution of industries with different factor concentration in the 1990s that: technology-intensive industries have substantially withered in the west regions and north central regions; labor-intensive industries have withered fully in northwest, southwest and north central regions, but achieved substantial growth in the south coastal and south central regions; capital-intensive industries, mainly those for raw materials and heavy machining, tend to move to the northwest and north central regions, such changes can be attributed to resource and market factors, as well as the readjustment by the government.

Further, we should set in a scientific way the industrial development target and distribution in different regions starting from their actual situation and according to the overall distribution of the state for industrial development. On one hand, we should keep a foothold on reality, to bring into maximum play the advantages of existing industries and make the local advantageous industries big and strong. Ivory-towered blind development of internationally advantageous industries without regarding of the restriction of objective conditions will result in punishment by economic laws; on the other hand, we should aim at development, integrate various production factors in the region, and even borrow factors outside the region that can be used, to develop promising merging industries and high and new technology industries, and foster the competitive advantages of industries. Nationwide, the general target is to promote the optimization and upgrading of industrial structure and continually upgrade the industrial competitiveness of the country.

Finally, we should follow the laws of development of modern industries, to build up competitive advantage of industries with the industrial "clustering" effect. "Clustering" refers to the "existence of a group of mutually associated companies, suppliers, associated industries and specialized institutions and associations in a special field of a specific area"[8] Researches by Michael Porter show that clustering has the functions of reducing transaction cost, increasing efficiency, improving stimulation method and creating collective wealth of information, specialized institutional system and reputation, and in such cases, many competitive advantages come from outside the company, that is, from the area and industrial cluster where the company is located. Therefore, government should change the ways of thinking, and improve the environmental for production growth mainly via policies, to promote the rise of industrial clusters. In the meantime, the principle of "diamond" model can be borrowed, and advantageous industries can be fostered through the understanding of production factors, demand conditions, associated industries and supporting industries and competition situation of the market, as well as the construction and optimization of this system, to build competitive advantage of the country.

3.3.3 Enterprise Competitiveness Based on Comprehensive Advantages of Large Country

Enterprises, instead of the country, are standing at the most front of international competition, just as Michael Porter said, "the performance of the country is associated with the competitiveness of enterprises, however, enterprises need to take initiative to search and apply important national resources related to enterprise competitiveness, and such important resources are never easily available as low-cost production factors. In creating competitive advantage, an enterprise able to become internationally successful should never be a passive spectator".[9] The most important to be done by an enterprise, is to enhance the enterprise competitiveness by relying on the CAOLC.

In the framework of CAOLC, diversification of main players is one of the basic characteristics. That is, a developing Large Country in transformation as ours has provided enterprises of various types with opportunities of survival and development. Therefore, enterprises should be good at utilizing CAOLC, to look for the "ecological position" suitable to it in the space of economic development. Business entities of different types also have their respective advantages, which are suitable to different industries, regions and operation and organization patterns. With respect to enterprise scale, big enterprises can easily gain advantages of scale, technologies and marketing, thereby forming internalization and resource integration advantages, which make it possible for them to obtain high monopoly profits; small and medium-sized enterprises have the advantages of flexible mechanism, the advantages to adapt to small sized market and technologies and the advantages of pioneering investment, suitable not only to developing traditional and featured industries in small cities and towns, but also to pioneering investment in high-tech parks of cities. Clusters of small and medium-sized enterprises formed in given regions can not only bring into play the advantages of small and medium-sized enterprises, but also borrow the advantages of big enterprises, to create a new type organizational structure with the advantages of both. They can gain relatively low cost and price advantages by replying on specialized division of labor and mutual cooperation, to increase the product competitiveness and market share. With respect to the enterprise ownership systems, the national conditions of China in the transformation period have provided enterprises of various ownerships with good environment of development, state-owned enterprises can adapt to market in reform by relying on the advantages of traditional industries, and the reputation, supplier and seller network established over a long time; and privately run enterprises can succeed in competition by relying on the advantages of clearly defined property right, flexible operation mechanism and low transaction expenses.

The characteristics of China of vast territory and rich resources can provide opportunities for enterprises to develop various industries, the important responsibilities of entrepreneurs are creating advantages of enterprises with the resource advantages by better allocating resources via the foundation role of market, to continually upgrade the competitiveness of enterprises, in terms of natural resources, China has two obvious advantages: one is large gross quantity and varieties, true to the saying "vast land and abundant resources". China ranks at the third place in the world in terms of territorial area, its land, forest land, grassland and fresh water resources all rank at the front places in the world, it has discovered 162 types of mineral resources, with the reserves of tungsten, antimony, vanadium and fluorite ranking at the front places in the world, with diversified types of biological resources, and fairly rich reserves of petroleum and natural gas. Second, there are great differences in space distribution, with strong complementarity in resources. In different geographic regions, the conditions of resources have distinctive features, with advantages in cultivated land, forest land and biological resources in the northeast, in coal, iron and lead and other mineral resources in North China, in the mineral reserves in northwest, and it is quite complete when all these in the whole country are put together. Such resource characteristics provide a foundation for division of labor in regions and in industries, so that enterprises in China can effectively integrate resources nationwide, to create comprehensive values. The features of large population and broad market of China provide enterprises with opportunities to obtain scale efficiency. China has a population of 1.3 billion, with an especially high proportion of rural population, containing huge consuming potential. Furthermore, demands of consumers are hierarchic, and people in different areas, at different ages and with different cultural levels and economic conditions have different demands for products. This broad market and complicated consuming structure provide unlimited space of market selection for Chinese enterprises, so that they can select market segments not only at different demand levels, but also on production links. So long as we really understand the consumers of China, our enterprises can make any product a big market. In the meantime, the scale of market expands the production scale, to form economies of scale and efficiency of scale, creating important conditions of sustainable development for Chinese enterprises.

3.3.4 Foreign Trade Competitiveness Based on Comprehensive Advantages of Large Country

Foreign trade is an important component part of the economy of a country, and is also the important content and external manifestation of the economic competitiveness of a country. The so-called "foreign trade competitiveness" is the ability of the tradable products, industries and enterprises of a country to open up and take up market in foreign countries, including the three levels of products, industries and enterprise competitiveness. It is the product competitiveness when seen from the outside, but hiding behind is the industrial competitiveness and enterprises competitiveness. For a late-developing Large Country like China, it must have characteristics different from those of other countries in the world when participating in international division of labor, and in summary, it is the "comprehensive advantages of Large Country", which is the real source of foreign trade competitiveness of China.

Traditional trade theory and modern trade theory have different understanding on foreign trade competitiveness: the traditional comparative advantage theory advocates allocating the resources of the country to enterprises with relatively high productivity, thereby establishing the trade on the difference of labor productivity between different countries; Michael Porter put forth a "diamond model" for obtaining competitive advantage on international market, thereby building the international trade on the foundation of fostering international competitive advantage. Both views contain reasonable contents, however, for the question of how comparative advantage theory and competitive advantage theory should be applied in a late-developing Large Country like China, we should put forth more clear thinking and countermeasures on the basis of in-depth research and understanding of the realities of China.

From a view of practice, the comparative advantage of China today still lies in labor intensive industries, and negligence of the existence of massive cheap labor in the country will be detrimental to the development of economy; if we do not endeavor to develop technology-intensive industries, China will never become an economically powerful country. The best way to solve this contradiction is to properly combine giving play to comparative advantage and fostering competitive advantage starting from the reality of comprehensive advantages of Large Country, with due consideration to the development of labor-intensive and technology-intensive industries. On foreign trade, we should not only give play to comparative advantage by developing the export of labor-intensive products, but also foster competitive advantage by developing the export of technology-intensive product, and the foundation for this combination is the comprehensive advantages of Large Country.

Now we analyze the advantages of export industrial or products, mainly using the trade competitiveness index method. This index is the ratio of the net export of a given industry or product to the total volume of its import and export, and the calculation formula is: $NTB_i = (X_i - M_i)/(X_i + M_i)$. Where X and M respectively stand for export volume and import volume, and i stands for a country of an industry or product. This formula is used mainly to measure comparative advantage, to further embody the competitive advantage and comprehensive advantages of Large Country, it is necessary to introduce the relevant factors of comprehensive advantages of Large Country into the foreign trade competitiveness measuring model and indicator system, thereby better describe the foreign trade competitiveness of a big developing country.

To deepen the research of foreign trade competitiveness based on CAOLC, we should choose three main points in our exploration: first, well combining comparative advantage with competitive advantage, unify them in CAOLC, and systematically analyze the basic characteristics of CAOLC, to build up the CAOLC theoretical system; second, we should tightly combine CAOLC with foreign trade competitiveness, to build up the foreign trade competitiveness system of China on the basis of CAOLC, further complete the existing foreign trade competitiveness index system and build corresponding models; and third, we should tightly combine CAOLC with the foreign trade development strategy of China, and base ourselves on the theory of CAOLC, so that our foreign trade development strategy become more clear, and detailed countermeasures and concepts are proposed.

As determined by CAOLC, we should establish a complete foreign trade system, which should embody the consistency of giving play to comparative advantage with fostering competitive advantage, and can realize sustainable development in the merging of international economy; and it should reflect the transformation from developing countries to developed countries, and gradually go to optimization of export commodity structure and upgrade trade patterns. Table 3.3 shows the foreign trade export commodity composition of China during 2003–2007.

This table clearly shows the foreign trade characteristics of China as a late-developing Large Country: first, it is a "big and complete" foreign trade commodity structure, which enables well utilizing the existing natural resources and economic and technological resources of China, and is conducive to building a complete foreign trade system, not separating from the realistic foundation of CAOLC; second, this is a structure demanding upgrading, the existing problems at present is: in the export of finished products from China, machinery and transport equipment (SITC7), miscellaneous products (SITC8) and finished products classified by raw materials (SITC6) constitute the largest portion, especially, the export of textile and metal products still account for about 20 %, indicating that the industrial products of China are still dominated by labor-intensive and low technology products. Although mechanical equipment representing the modern industrial level take a fairly large share in the export commodities of China, FDI and processing trade play a dominating role in them, while the export of high-tech products from

Table 3.3 Export commodity composition during 2003–2007. *Unit* 100 m USD, %

	2003		2004		2005		2006		2007	
Commodity composition (classified as per SITC)	Amount	Prop.	Amount	Prop.	Amount	Prop.	Amount	Prop.	Amount	Prop.
Total value	4382.27	100.00	5933.25	100.00	7620	100.00	9689.37	100.00	12,180.1	100.00
I. Primary product	348.11	7.94	405.49	6.83	490.40	6.43	529.19	5.46	615.50	5.05
Category 0	175.31	4.00	188.64	3.18	224.80	2.95	257.23	2.65	307.50	2.50
Category 1	10.19	0.23	12.14	0.20	11.80	0.15	11.93	0.12	14.00	0.11
Category 2	50.32	1.15	58.43	0.98	74.90	0.98	78.60	0.81	91.50	0.75
Category 3	111.14	2.54	144.80	2.44	176.20	2.31	177.70	1.83	199.40	1.63
Category 4	1.15	0.03	1.48	0.02	2.70	0.01	3.73	0.04	3.00	1.50
II. Industrial product	4034.16	92.06	5527.76	93.17	7129.60	93.57	9160.18	94.54	11,564.70	94.94
Category 5	195.81	4.47	263.60	4.44	357.7	4.69	445.30	4.60	603.60	4.95
Category 6	690.18	15.75	1006.46	16.96	1291.30	16.95	1748.16	18.04	2198.90	18.05
Category 7	1877.73	42.85	2682.60	45.21	3522.60	46.23	4563.43	47.10	5771.90	47.38
Category 8	1260.88	28.77	1563.98	26.36	1941.90	25.48	2380.14	24.56	2968.50	24.37
Category 9	9.56	0.22	11.12	0.19	16.10	0.21	23.15	0.24	21.80	0.18

Source Calculated from relevant data in *China Commerce Yearbook* for various periods

local enterprises of China is still quire far from the world average level. The total import and export trade volume of China has exceeded 2.56 trillion USD, and this massive foreign trade scale means that China can exert influence on some important commodities on the international market. However, to go from a Large Country economy to a powerful country economy, we must get competitive advantage on the international market, and this requires us to make breakthroughs in critical industries and technologies, and support and develop the export from a batch of critical industries with fair competitiveness, to obtain more added value and gradually form the competitive advantage of the country.

The economic development strategy based on CAOLC and the corresponding industrial development strategy, enterprise development strategy and foreign trade development strategy, are the dominant strategy suitable to the national conditions of China. It contains the complicated national conditions of China with broad connotations, and absorbed the advantages of a number of economic development strategies, therefore it can realize the proper unification of giving play to comparative advantage and fostering competitive advantage, and the unification of creating competitive advantage and giving play to later-mover advantage.

Notes

1. Simon Kuznets is the first foreign scholar defining Large Countries and small countries quantitatively, setting the initial criteria of classification. Refer to: Simon Kuznets: *Economic Growth of Nations*, the Commercial Press 1985 Edition.
2. China Modernization Strategy Research Subject Group: *Regional Modernization in India*, China Net Feb. 20, 2004.
3. Xu Chengde: *Policies Supporting the Development of Modern Agriculture in Brazil*, *Modern Agricultural Science and Technology* 2009 Vol. 6.
4. Liu Zhiyong, Hu Yongyuan: *Empirical Test of Total Factor Productivity, Capital Accumulation and Regional Differences*, *Statistics and Decision-making* 2008 Vol. 12.
5. Huang Lixin: *Economic Recession in the United States: Causes and Its Impact on World Economy*, *Internal Circulation* 2002 Vol. 1.
6. Hai Wen: *China's Economy Has Advantages of Large Countries and Can Grow Powerfully for Further 20 Years*, *Beijing Morning Post* June 1, 2007.
7. Published by the World Bank: *2009 World Development Report*, Tsinghua University Press 2009 Edition.
8. [US] Michael Porter: *The Competitive Advantage of Nations*, Huaxia Publishing House 2002 Edition.
9. Same as [US] Michael Porter: *The Competitive Advantage of Nations*, Huaxia Publishing House 2002 Edition.

Chapter 4
Model: Analysis of Formation Mechanism

In the previous chapter, the concept of "comprehensive advantages of Large Country" was put forth, and theoretical interpretation was made, with emphasis on analyzing its specific performance in the industrial competitiveness, enterprise competitiveness and foreign trade competitiveness. This chapter will build a model to describe the formation of CAOLC and its functioning mechanism, and also analyze in detail the forming route of CAOLC and the mechanism to convert it into competitive advantage and promote industrial upgrading, to deepen the cognition and understanding of CAOLC.

4.1 Magic Gourd Hypothesis

In studying the outbound investment of China in 2006, the author proposed the model for analyzing the forming route of CAOLC, with the basic thinking that, the characteristics of regional diversification, economic diversification and technological diversification in late-developing Large Countries will lead to main player diversification, motivation diversification, industrial diversification and advantage diversification in outbound investment, thereby forming a comprehensive economic efficiency. This model is based on the national conditions of China, with "diversification" as the link, to describe the forming mechanism of CAOLC, and it is named as "Magic Gourd theory" after its shape.[1] As shown in Fig. 4.1.

In recent two years, the author conducted research of CAOLC as a general phenomenon in the economic development in Large Countries, without rigidly adhering to the particularities of China. This has achieved new cognition and understanding of the forming mechanism of CAOLC, thereby building a new "Magic Gourd model", as shown in Fig. 4.2.

It can be seen in Fig. 4.2 that, normally, a Large Country has the special conditions of rich natural resources, vast national territory, large population and high market potential, and the derived scale, differences, diversity and completeness of

Y. Ouyang, *The Development of BRIC and the Large Country Advantage*,
DOI 10.1007/978-981-10-0633-3_4

后发大国 Late developing Large Countries
地区多元化 Regional diversification
经济多元化 Economic diversification
技术多元化 Technological diversification
对外直接投资 Out-bound direct investment
优势多元化 Advantage diversification
动机多元化 Motivation diversification
主体多元化 Main player diversification
产业多元化 Industry diversification
区域多元化 Regional diversification
大国综合优势 Comprehensive advantages of Large Country

Fig. 4.1 Formation of comprehensive advantages of large country—magic gourd theory

大国 Large Country
自然资源 Natural resources 国土面积 Territorial area 人口数量 Population size 市场潜力 Market potential
规模性 Scale 差异性 Differences 多元性 Diversity完整性 Completeness
分工优势 Division of labor advantage 互补优势 Complementation advantage适应优势 Adaptation advantage 稳定优势 Stability advantage
大国综合优势 Comprehensive advantages of Large Country

Fig. 4.2 New magic gourd model: formation of comprehensive advantages of Large Country

economy, and further forming the advantages of division of labor, complementarity, adaptation and stability, and this is the basic route of forming the CAOLC.

(1) The advantages derived from scale of economy: a country of big scale has big scale of economy and industries, and this feature can form some advantages. First, the scale of economy can deepen the division of labor, to realize specialized production, upgrade the technological and management level and improve the product quality and production efficiency. Second, the scale of economy can cut cost, with the expansion of production scale, the cost tends to reduce within the range of marginal benefit, therefore increasing profit or reducing product prices. Third, the scale of economy can form pillar industries, the expansion of economic scale will enlarge the scale of industries, thereby forming some pillar industries playing important roles in the national economy (Fig. 4.3).
(2) Advantages derived from economic differences: with a vast national territorial area, there will be great differences between different regions, with different advantages of resources, industries and products, and in the viewpoint of a country, it can form complementarity advantages of regions, industries and products. First, differences can lead to advantage complementation between regions, in a Large Country, different areas have different natural and

规模性 Scale
分工深化优势 Advantage from further division of labor 成本减少优势 Advantage of reducing cost 支柱产业优势 Advantage of pillar industries

Fig. 4.3 Advantages derived from scale

economic characteristics, it is suitable to develop industries in some regions, to develop agriculture in some regions, and to develop tourism in some other regions, thereby forming various economic regions with respective main sectors, and the regions can complement with each other in advantages and develop in a coordinated manner. Second, differences lead to advantage complementation between industries, a country needs industries on various aspects, but only a Large Country has the required conditions to develop these industries in an all-round way, so that the industries of machine building, electronic, chemical, aerospace, biology and textile can complement with each other in advantages. Third, differences lead to advantage complementation between products, the citizens in a country are at different levels, forming different groups of consumers, and they have different demands for products, therefore, products of different levels should be made to meet their demands, the feature of rich and complete range of product varieties in a Large Country can form the advantage of product complementation, and in terms of foreign trade, it can also meet the demands from different countries and levels (Fig. 4.4).

(3) Advantages derived from economic diversity: Large Countries usually have a multi-structure, with the characteristics of diversity on human resources, technologies and products, and such diversified characteristics can adapt to different demands in production and life, thereby forming an adaptability advantage. First, diversity leads to adaptability of human resources, Large Countries have rich human resources, with professionals in different fields and at different levels (some are suitable to industrial production, some suitable to agricultural work and some suitable to work in tertiary industry), to enable coordinated development of these different fields. Second, diversity leads to adaptability of technologies, professional technologies are also divided into different levels, Large Countries have both applicable technologies and high and new technologies, and this diversified technological structure can better meet the demands in the development of national economy. Third, diversity leads to adaptability of products. Consumers have diversified demands for

差异性 Differences
区域互补优势 Complementation advantage of regions 产业互补优势 Complementation advantage of industries 产品互补优势 Complementation advantage of products

Fig. 4.4 Advantages derived from differences

多元性 Diversity
人力资本适应性 Adaptability of human resources 技术适应性 Adaptability of technologies 产品适应性 Adaptability of products

Fig. 4.5 Advantages derived from diversity

完整性 Completeness
产业稳定性 Stability of industries 产品稳定性 Stability of products 就业稳定性 Stability of employment

Fig. 4.6 Advantages derived from stability

products, some need automobiles and motorcycles, some need bicycles, some need running machines and some need bowling, the diversified products in Large Countries can meet this demand structure of consumers (Fig. 4.5).

(4) Advantages derived from economic completeness: Large Countries usually have complete economic systems, a full range of industrial sectors and strong ability to resist against external risks, therefore the internal circulating system can be relied to maintain economic stability. First, completeness leads to industrial stability, Large Countries have complete range of industrial sectors, and their domestic industries can support the development of economy, therefore, industries are not subjected to destructive impact by external factors, and they can maintain long-term stability. Second, completeness leads to stability of products, the industrial sectors can make products in different fields, and human resources and professional technologies at different levels can make products of different levels, therefore, they can meet the demands from different consumers and maintain long-term stability of domestic market. Third, stability can lead to stability of employment, the market demands and product demands at diversified levels can maintain economic prosperity, thereby producing jobs in various areas, to keep steady employment in the society and ensure that people can live and work in peace and contentment (Fig. 4.6).

4.2 Formation of Comprehensive Advantages of Large Countries

1. Effect of scale and division of labor

The comprehensive advantages of Large Country are formed mainly by the economies of scale, differences, multi-structure and independent systems originated from the "big size", and the thereby derived advantages in division of labor, complementarity, adaptability and stability.

The basic conditions of economy of scale are the corresponding scales of factors, including the scale of labor, of resources, of capital and of market. If the production factors in a country are in too small scale, they cannot form economy of scale in a sector, so the production activities in that country can only be in a state of diseconomies of scale. The international trade model proposed by Krugman indicates that: the varieties of industries or products in a country depend on the scale of production factors and production scale of products in that country, and the production scale of products is actually the setup cost of the products. Economies of scale are inevitably linked with division of labor, and it is the fundamental motivation and condition leading to division of labor. Large Countries have large scale of production factors, and can concentrate resources of certain quantities, furthermore, the domestic demand is high and domestic market is vast, leading to division of labor and specialized production, pushing forward technological segmentation and progress. The economic scale characteristics of Large Countries make the division of labor both within regions and in the country quite developed, thereby forming scale advantage and advantage in division of labor.

2. Differences and complementarity

A Large Country has a vast territory, and different regions usually have differences, including differences on factor endowment and industrial structures. Different regions have different natural resources, human resources, technological conditions and social and economic conditions, with their respectively features, and some of these conditions are their comparative advantages. The theory of Heckscher–Ohlin has proved that, under the conditions of free trade, all countries or regions tend to export products made by concentrated use of their relatively abundant factors, and import products made by concentrated use of factors they are in relative shortage, and this is the principle of modern comparative advantage. In the production, if all regions select industries, products and technologies according to the comparative advantages of factors, such economies are competitive, and the factor endowment structure will upgrade fairly quickly, thereby conducive to the sustainable development of economy. Therefore, in a Large Country, economic development strategies should be selected in regions with differences according to the comparative advantage of factors: in regions relatively rich in labor resource, industries with relatively intensive labor force should be developed in priority; in regions relatively rich in capital, industries with relatively intensive capital be developed in priority; and in regions relatively developed in technologies, technology-intensive industries be developed in priority. In this way, the industries in all regions will have relative competitiveness, and can obtain the maximum efficiency. In the whole country, all regions can tend to form factor complementation, industry complementation and economy complementation, and this is the complementarity advantage.

3. Heterogeneity and adaptability

Late developing Large Countries generally have multi-structures, including technological multi-structure, human resources multi-structure, and economic multi-structure. The "dual economy theory" put forth by Arthur Lewis is in fact a "dual-technology theory", including advanced science and technology and technology-intensive, traditional handicraft and labor-intensive types. To go further, "multiple structure" can be seen. The multiple structure of technologies includes high and new technologies, applicable technology and traditional handicraft technologies; and the multiple structure of human resources includes high-tech professionals, professionals for applicable technologies and for traditional handicraft technologies. In Large Country economy, all these technologies and professionals are needed in economic development, and they can suit the production in different industrial sectors and for different products. The key point is that we cannot simply judge the "multiple structure" as superior or inferior, instead, we should see the adaptability feature of the "multiple structure", that is, different technologies and human resources are suitable to the production in different industries and for different products. Provided that the structure is rational, all these different industries are needed in the development of national economy, and different products are also needed in the life of the people. In Large Country economy, it is the "multiple structure" of technologies and human resources and their adaptability features that support the development of different industries and the production of different products, thereby meeting the diversified demands in the development of national economy and by the people, promoting the harmonious development of Large Country economy.

4. Independence and stability

The development of Large Country economy has an important feature that the completeness and independence of the economy system should be considered strategically, and usually a fairly independent system of national economy and an industrial system with complete range of varieties should be established. The development of Large Country economy involves quite large amount of production and capital demand, the consumption of most products cannot be fully relied upon foreign trade, and the accumulation of most capital cannot be fully relied upon the introduction of foreign investment. Therefore, Large Countries usually are less dependent on foreign trade than small countries. Kuznets has found the reversed relationship between country scale and trade dependence, and Chenery's research also proved this relationship. The dependence on foreign trade of medium and small countries and regions is much higher than that of Large Countries, that is, the economic growth of medium and small countries and regions depends more on the expansion of export. The economic independence feature of Large Countries enables them to establish internal circulating systems, thereby they have the self-regulation ability, to guarantee the stability of economic system and form an advantage of stability. In an international financial crisis, it usually demonstrates the "Large Country effect" of stable finance and stable economy.

The CAOLC theory reveals the realistic characteristics of late-developing Large Countries, and well compromises the debate between "comparative advantage" and "competitive advantage", so that the roles of "comparative advantage" and "competitive advantage" can be well combined, to formulate scientific and rational foreign economic relation and trade strategies. Longitudinally, China is in the transition period from a pattern with dominant "comparative advantage" to that with dominant "competitive advantage", therefore we should attach importance to both giving play to comparative advantage and fostering competitive advantage; transversally, China is in a situation with coexisting "comparative advantage" and "competitive advantage", therefore some regions, industries or enterprises should attach importance to giving play to "comparative advantage", while some others should attach importance to giving play to "competitive advantage". The proposing and application of the CAOLC theory is the result of researching the special national conditions and economic development strategies of late-developing Large Countries with materialistic dialectics.

The diversification of regions, economy and technologies can result in comprehensive advantages in the viewpoint of integrating resources, and it can further lead to diversification of development advantages, diversification of motivations, diversification of main players and diversification of industries. Realizing this "diversification" strategy can obtain comprehensive economic efficiency. The economic development strategy based on CAOLC and the corresponding industrial development strategy, enterprise development strategy and foreign trade development strategy, are the advantage strategy suitable to the national conditions of late-developing Large Countries, and it has incorporated the advantages of a number of economic development strategies with its broad connotation. It can accommodate the development of both comparative advantage industries and those with competitive advantage, thereby realizing the organic unification of comparative advantage and competitive advantage; some regions can implement "gradient transfer" strategy, and some regions can implement "reversed gradient transfer" strategy, thereby realizing the organic unification of "gradient transfer" and "reversed gradient transfer".

4.3 Mechanism to Convert Comprehensive Advantages of Large Country into Competitive Advantages

On the international market, CAOLC only indicates a potential possibility to obtain comprehensive benefits, while comprehensive benefits can only be obtained by realizing the values of products. Economic entities can obtain more comprehensive benefits only by converting the comprehensive advantages into competitive advantage. Late-developing Large Countries should make full use of the CAOLC, i.e. the later-mover advantage[2] and potential early-mover advantages[3] of developing countries, while avoiding the later-mover disadvantage of developing countries.

The later-mover advantages of developing countries include advantages on technologies, capital, labor, institutional system and structure, among them, the later-mover advantage of technology is contained in the technology difference between late-developing countries and early-mover countries, the bigger difference, the higher potential energy; by institutional later-mover advantage, it means that late-developing countries learn from the early-mover countries on their institutional systems, that is, imitating and follow various advanced systems and localizing them, to get efficiency and benefits. It is a potential relative advantage, and can play its role on the basis of certain social ability; it is a process of continual repeated learning, accumulation and creation, and also a process of continual imitation and innovation. The potential early-mover advantages of late-developing Large Countries are the early-mover advantages on factors, technologies and institutional system of developing countries after some development in the international economic and political games, but they are not resulted from action or non-action of relevant countries. Developing countries can obtain later-mover advantage through technological and institutional imitation, however, as developing countries should reform their institutional system and offend some vested interests, therefore they may tend to technological imitation while neglecting institutional imitation. Although this can enable them to achieve very good development in short term, it may leave many hidden perils for the long-term development, and even lead to failure of long-term development, and this is the "late-mover disadvantages" of late-developing countries.

Then, how can a country bring into play its CAOLC? First, the whole society should be able to make correct response to comprehensive advantages, to know what the existing comparative advantage industries are in the country, what the potential comparative advantage industries (capital-intensive industries) are, where the existing later-mover advantage is embodied, where the early-mover advantages are embodied, and how the production factors be allocated between industries to embody these two advantages. In implementing the strategy of CAOLC, the most important is to prevent trapping the national economy into an inertia because of comparative advantage, as a result, the economic structure cannot be regulated in good time when the market is not functioning well, thereby stagnating the development of national economy; in the meantime, negative effect should be prevented that the resources allocation is seriously distorted and market allocation can hardly play its role because of the implementation of catch-up strategy.

In our framework, countries adopting comprehensive advantage development strategy will import products without comparative advantage, and at the same time export products with comparative advantage and potential comparative advantage. In the domestic and overseas market competition, such industries with potential comparative advantage can continually upgrade their own technological abilities because of the effect of technological diffusion and self-learning, therefore their potential comparative advantage can be converted into realistic comparative advantage. Upon gaining the realistic comparative advantage, enterprises will expand their scale of production, and attract more capital investment and companies to join it, to increase the scale of economy and industrial concentration, and further to form an industrial chain.

In the meantime, as a late-developing Large Country, we have very high technological standard in some industries or some fields, such as aerospace technology, information technology, biotechnology and new energy automobile technology. These technologies are at internationally advanced positions, and can be developed as early-mover advantages. After the impact of international financial crisis, people came to know the important significance of the new industrial revolution triggered by the new scientific and technological revolution to promoting the sustainable development of economy. To follow this trend, the Chinese Government made the decision to develop strategic emerging industries, and new energy, energy conservation and environmental protection, new energy automobile, new materials, the next generation of IT technology, biotechnology and high-end equipment manufacturer were selected as strategic emerging industries for fostering and development, actually, this is to make use of the "early-mover advantages" of China in these emerging industries, to gain competitive advantage by further fostering them.

As determined by the "later-mover advantage paradox"[4], an economic system, while making effective use of later-mover advantage, must pay attention to building and fostering early-mover advantage. It is because, with the development of economy, later-mover advantage will gradually get lost, and the economy of the country will be trapped in a stagnant state, therefore, late-developing countries, especially late developing Large Countries, should build up some early-mover advantages in advance, otherwise, it will be difficult to realize self-reliance and surpassing development by only making use of the current later-mover advantage without attaching importance to the future sustained development to go to the phase of early-mover advantage.

Therefore in short term, developing countries can adopt appropriate industrial policies and trade policies to support and protect on a selective and temporary basis their own industries still in the primary stage of development, so that their infant industries can enjoy the benefits of economies of scale as well as learning by doing; and in the long run, developing countries should endeavor to increase their own human resources and their own technological ability and production efficiency, so as to determine and consolidate the comparative advantage fostered in the primary stage of industrial development with protection and support. Developing countries can steer the comparative advantage in their favored direction only by a portfolio combining short-term and long-term policies, so that their industrial structure can continually grow and expand to advanced level, to realize the upgrading and transformation of their own industrial structure.

4.4 Mechanism of Comprehensive Advantages of Large Country Promoting Industrial Structure Upgrading

The conversion of industrial structure from a resources-intensive and labor-intensive one to a capital-intensive and technology-intensive industries, as well as the upgrading from low end to high end within industries, are critical in the transformation of economic growth pattern. The comparative advantage development theory states that, adopting the comparative advantage development strategy can achieve the highest speed of capital accumulation, thereby upgrading the factor endowment structure, and finally upgrading the industrial structure, therefore, governments of under-developed countries should aim at promoting the structural upgrading of factor endowment, instead of upgrading the industrial and technological structures, because once the factor endowment structure has been upgraded, profit motivation and competition pressure will drive enterprises to spontaneously upgrade their technological and industrial structures.[5] However, in the opinions of the author, this development strategy cannot effectively upgrade the industrial structure of Large Countries. China has adopted comparative advantage strategy over a long time, and obtained fairly big economic surplus, however, it is also accompanied with deterioration of trading conditions and growth of immiserization. With changes in the international and domestic economic situation, some unsuitability of the comparative advantage strategy has appeared in China, mainly because a Large Country will encounter problems different from those in small countries when comparative advantage is adopted to develop economy, thereby resulting in difficulties in the smooth upgrading of industries and in development of economy, mainly in the following two aspects:

(1) Deterioration of trade conditions of Large Countries and factor reward diminishing anathema. There is a very big difference between Large Countries and small countries: Large Countries can govern the prices of products on international market, while small countries can only accept prices on international market. The prices formed by low-price competition among the labor intensive industries in late developing Large Countries are very close to cost, with only limited space of price adjustment. However, to developing small countries, even if the trade conditions deteriorate, they can mitigate the detrimental effect produced by deterioration of trade conditions of some categories of commodities by regulating the export structure and diversified production, but it is much more difficult for Large Countries to regulate export structure than small countries, because Large Countries can influence and even determine the international price in quite many industries, and large total resource quantities are required for Large Countries to upgrade their industrial structure. According to typical neoclassical growth theory, the rise of capital to labor ratio reduces the marginal contribution of capital inventory to output, therefore growth is subjected to restriction by diminishing marginal return of factors. Large Countries are more inclined to the "anathema of diminishing

marginal return of factors", because as long as a Large Country increases the supply of one product, the price of that product on world market will be decreased. Therefore, the increase of output in Large Countries will reduce the value of output through the deterioration of trade conditions, amplifying the effect of factor accumulated diminishing marginal return.[6] Also, it is far more difficult for a developing Large Country than a developing small country to get rid of the "comparative advantage trap" and "immiserizing growth" characterized by long-term low-added value and severe trade conditions. To China, the trade conditions have been in the trend of deterioration since the 1980s. Also, according to the view of "fallacy of composition theory": one country can be successful in adopting a strategy, but when many countries adopt the identical strategy, all countries will fail. When developing countries such as India, Brazil and Mexico with similar comparative advantage also implement free trade and actively participate in international division of labor and export industrial finished products in large quantities, it will be more difficult to improve the trade conditions for the labor-intensive industrial finished products, the competition among labor-intensive products will become daily severe, and the probability of fallacy of composition will increase substantially.

(2) Rigidity of industrial transformation and upgrading in Large Countries, according to the development strategy of comparative advantage, the upgrading of industrial structure depends on the dynamic changes of factor endowment structure (K/L): Prof. Justin Yifu Lin believed that if underdeveloped countries choose to follow the comparative advantage strategy, the factor endowment structure upgrading will be very fast. The four tigers in East Asia, and even Japan in bigger scale, smoothly realized upgrading of industrial structure by improving their factor endowment structure.[7] However, China, as a developing Large Country, has its special conditions, it is lacking in capital but has abundant labor, and more important, due to the existence of the dual economic structure, with the industrial restructuring and completion of market economic system, massive labor force moved out of rural areas, so that the labor intensive industries with comparative advantage in China have almost unlimited supply of labor force, which makes it very difficult to improve the factor endowment in China in a short period of time. Just as pointed by Yu and Botao (2006), wages is the most important index to measure the factor endowment structure of capital to labor ratio, after the 1980s, there was no obvious increase in the actual wages of migrant rural workers in China, and this means that, after rapid development for nearly three decades in China, after massive accumulation of capital, the resource endowment in China has not been fundamentally changed, and this has resulted in difficulties in industrial structure upgrading in China. "Any inducement of production factor intensity transformation triggered by technological progress will be relentlessly disintegrated by the fact that China has massive inexpensive labor forever and there is extremely abundant labor factor. … the overwhelming structural form of labor intensive industries makes it difficult to increase labor productivity, resulting in the rigidity of transformation of industrial structure."[8]

Therefore, to a country with fairly small scale of economy, adopting the comparative advantage development strategy can undoubtedly achieve the highest speed of capital accumulation, thereby upgrading the factor endowment structure and finally upgrading the industrial structure. However, a Large Country with unbalanced economic and technological development cannot effectively upgrade its industrial structure with this development strategy, the "diversified economy" and factor endowment characteristics of a late-developing Large Country has determined that it should make use of the CAOLC to promote the upgrading of its industrial structure.

First, based on the economic scale characteristics, Large Countries can concentrate high quality resources to develop some key high end industries, to promote the upgrading of key industries. According to the analysis framework of comparative advantage theory, with countries as decision-making units, if two countries, one small and one big, have the identical factor endowment degree (K/L), then these two countries should follow the same route to regulate their industrial structures. But in fact, the production possibility curves of these two countries are not identical, because many industries have the access threshold of minimum factor endowment. The industrial structure of small countries will gradually transform from labor-intensive to capital-intensive and technology-intensive with the increase of the overall factor endowment degree. Different from small countries, Large Countries with comparatively large overall scale can concentrate part of resources to access into industries with higher factor endowment requirements, thereby greatly shortening the time of upgrading the whole industrial structure.

Next, based on economic multiple characteristics, Large Countries can make use the difference of regional factor endowment to expand the scope of selection of industries, to push forward the upgrading of industrial structure level by level. The developed east coastal regions of China, after three decades of reform and opening up, have accumulated large amount of material capital and human capital, the factor endowment has been changed greatly, and technological level also been greatly raised, with the ready conditions to develop capital-intensive and technology-intensive industries to a certain extent, therefore upgrading of industrial structure can be encouraged first in the developed east regions. Furthermore, between the east coastal regions and central and west regions, and between extra-large cities and small and medium-sized cities in China, there are not only great differences in their factors such as labor, capital (including human capital) and technology, their development extent and completion of market institutions and systems also differ greatly. Therefore, by following the theoretical standard of comparative advantage, there should be different targets and routes in these regions for the upgrading of industrial structure. If they all aim at developing capital-intensive and high and new technology industries, it is only grasping the shadow instead of the essence in the relative backward central and west regions and small and medium-sized cities. If all regions of the whole country all upgrade their trade structure with no regard to practical conditions, it will probably achieve little with great efforts.[9] Also, if we put the total resource endowment conditions of the whole country into labor intensive comparative advantage industries, the upgrading of industrial structure will surely be slowed down because of negligence of the economic diversity characteristics. Therefore, the upgrading of industrial structure in China must be founded on the reality of "diversified economy", and be pushed ahead level by level. For example, we can realize the space restructuring of advantages of the east region and central and west regions through division of labor between regions. In the division of labor system between industries, the east region can, at the appropriate time, transfer its labor intensive industries to the under-developed central and west regions, to develop in priority capital-intensive and technology-intensive industries; in the division of labor system within industries, the east region can give play to its advantages in capital and technologies, to mainly specialize in high quality products, while the central and west regions can bring into play their advantages on labor and land resources, to mainly specialize in ordinary products; in the division of labor system within products, the east region can be mainly specialized in the high end links of value chain such as research and development, marketing and management, while the central and west regions can mainly specialize in the labor-intensive processing sectors. With this space restructuring of comparative advantage and cooperation in industrial division of labor in different regions, it can be expected that the developed east coastal region can, through self-dependent innovation, develop capital intensive and technology intensive industries, and extend and even strengthen the traditional comparative advantage of China in the under-developed central and west regions through technological transfer and industrial transfer, and finally realize the "hierarchic upgrading" of the overall industrial structure in the whole country.

Finally, according to the autonomous feature of economy, Large Countries can adopt strategic industrial policies, to promote the industrial structure upgrading by fostering competitive advantages. Regional, economic and technological diversification can be deduced to diversification of development strategies, motivation, main players and industries, and this "diversified" economy characteristics are the basis of adopting strategic industrial policies to promote the upgrading of industrial structure. Large Countries with unbalanced economic and technological development need not go into international division of labor according to comparative advantage in a totally passive manner, instead, they should well combine the utilization of comparative advantage with fostering new competitive advantage, to actively obtain advantages through technological progress and human capital investment, so as to change the general comparative advantage in a country and promote the upgrading of industrial structure.[10] Government can, based on its CAOLC, support the capital-intensive industries in a short period of time, to promote them to improve technology, and obtain the benefit of economies of scale and learning by doing. Furthermore, the domestic market advantages of Large Countries can lay some foundation for the feasibility of their strategic industrial policies, thereby guaranteeing the effective optimization and upgrading of the industrial structure in the country. Historical experience shows that, some late-developing

Large Countries with fairly successful development of economy in history did not completely adopt comparative advantage in their economic development, while developing their own comparative advantage industries, they also endeavored to develop some medium and high end industries through self-reliance research and development, so they actually adopted a limited catch-up strategy.[11]

In short, Large Countries with unbalanced economic and technological development should make necessary correction to the comparative advantage development strategy according to the characteristics of Large Country and features of diversified economy, strengthen the research and development in labor intensive industries while giving full play to the comparative advantages of labor intensive industries, increase the added values of products, attach importance to cultivating the ability of self-reliance innovation and solve issues in critical technologies on a selective basis; they should make use of the features of Large Country to expand the scope to select industries, make use of the diversity of economy and technologies and differences in regional resource endowment, to push forward the upgrading of industrial structure by regions, by industries and by levels; and they should form new advantages of a Large Country through trans-regional resources cooperation and technological cooperation, to realize the organic combination of natural evolution of industries under the comparative advantage strategy with the self-reliance advancement under the catch-up strategy, to effectively promote the industrial structure upgrading in Large Countries.

Notes

1. Yao Ouyang: *Build up Comprehensive Advantages of Large Country*, *Journal of Hunan University of Commerce* 2006 Vol. 1.
2. Gerchenkronft was the first creating the theory of later-mover advantage, and he stated that later-mover advantage is extensively embodied in factors of capital, technologies and institutional systems. Refer to: Gerschenkron alexander, 1962, Economic Backwardness in Historical Perspecture, Harvard University Press. At present, developed countries account for 96 % of the total research and development expenditure in the world, and there is a huge technological gap in China from developed countries, therefore import and imitation of foreign technologies are the main route for China to utilize the later-mover advantage in technology, Chinese enterprises depend on foreign technologies by as high as over 50 %, and in core technologies and equipment in most important industries, such as 100 % of optical fiber manufacturing equipment and 80 % of IC chip manufacturing equipment and petrochemical equipment depend on import (refer to http://www.mof.gov.cn/qiyesi/zhengwuxinxi/di-aochayanjiu/200806), to these industries, utilization of later-mover advantage is an important means to increase their competitiveness.
3. Early-mover advantage is so termed relative to later-mover advantage, China can focus on self-reliance research and development in some sectors to reach the advanced level in the world, for example, substantial technological progress has been made in the electronic industry of China with computer, household electrical

appliances and communication, so in this sector, we can encourage self-reliance innovation to reach the world leading position, to realize early-mover advantage.
4. If a backward country cannot make effective use of later-mover advantage, it may possibly fall into the "later-mover advantage trap", which will restrict the further development of economy, resulting in difficulty to finally realize catch-up, and even further expanding the gap from advanced countries; this phenomenon is also referred to as "later-mover advantage paradox".
5. Justin Yifu Lin: *Development Strategy, Viability and Economic Convergence, Economics (Quarterly)* 2002 Vol. 2.
6. Acemoglu and Ventura, 2002. "the World Income Distribution". *Quarterly Journal of Economics*. vol. 117.
7. Justin Yifu Lin, Cai Fang and Li Zhou: *Comparative Advantage and Development Strategy—Reinterpretation of "East Asia Miracle", Social Sciences in China* 1999 Vol. 5.
8. Cui Yonglie: *Double-edged Sword of Trade and Development, Reading* 2005 Vol. 11.
9. Zhang Yabin and Yi Xianzhong: *Circling and Layering Upgrading of Trade Structure and Change of Foreign Trade Growth Pattern in Heterogeneous Large Countries, International Trade* 2007 Vol. 2.
10. Xu Jianbin and Yin Xiangshuo: *Deterioration of Trade Conditions and Effectiveness of Comparative Advantage Strategy, World Economy* 2002 Vol. 1.
11. Yang Rudai and Yao Yang: *Limited Catch-up and Development of Large Country Economy*, essay at 2006 Sixth Annual Conference of China Economics.

Chapter 5
Framework: Factor Structure Analysis

In the previous chapters, we described the formation and acting mechanism of comprehensive advantages of Large Country, in this chapter, we will further explore on how the comprehensive advantages of Large Country, i.e. natural resources, financial capital, human capital, technological progress, market potentials and foreign trade play their roles by influencing the relevant factors in economic growth, and conduct special research on the important role of institutional innovation in bringing into play the comprehensive advantages of Large Country, so as to provide a fairly complete framework for analyzing the comprehensive advantages of Large Country.

5.1 Determinants of a Large Country's Economic Growth

Economic growth refers to the increase of national output. Generally speaking, in the research of economic growth, consideration should be given to various interpreting variables including the main determinants for expansion of production capacity or economic growth. People came to understand the origin of economic growth with a progressive process, from taking physical capital accumulation as the determinant of economic growth to attaching importance to technological progress and emphasizing decisive role of human capital accumulation and institutional factors in economic growth, and today, a fairly complete economic growth factor analysis system has been formed.

Classical economists attributed economic growth mainly to capital accumulation. Adam Smith's book *An Inquiry into the Nature and Causes of the Wealth of Nations* is in fact a monumental work on researches of economic growth. He stated that saving is an important source of capital increase. With certain capital

Y. Ouyang, *The Development of BRIC and the Large Country Advantage*,
DOI 10.1007/978-981-10-0633-3_5

accumulation in an appropriate scale of market, the labor productivity and profit rate can be increased through division of labor, and capital accumulation can be increased, thereby realizing continual increase of national wealth; the great geographic discovery, industrial revolution and application and improvement of new technologies of transportation and communication can continually open up new markets and expand foreign trade, to strengthen the momentum of national wealth growth.[1] It can be seen that the originator of economics Adam Smith has provided us with an analysis framework with capital accumulation as foundation, and including such determinants of natural resources, technological progress, market size and foreign trade. Subsequently, the Harrod-Domar model, with the prerequisite to assume that both the ratio of production capital to labor and the technical innovation level remain unchanged, also came to the conclusion that economic growth rate depends on saving rate, believing that capital accumulation is a decisive determinant in continual economic growth.

The neoclassical economics developed the theory about economic growth determinants, Solow model concluded that economic growth rate is determined by the three main factors of technological progress rate, capital growth rate and labor growth rate, mentioning for the first time the decisive role of technological progress in economic growth, and forming the analysis framework including technology, capital and labor, Romer and Lucas stated that, the long-term economic growth can be explained by endogenous technological progress or human capital accumulation. With them, physical capital accumulation is no longer the motive force for economic growth, and it is the progress in knowledge, research and development and human capital that determine the economic development. The new growth theory formed in the 1980s mainly achieved progress in three aspects: the first is adding the innovation theory represented by technologies and knowledge; the second is incorporating human capital into the scope of capital; and the third is that the analysis of institutional factors has entered the field of research.

Among contemporary economists, Arthur Lewis and Simon Kuznets put forth the systematic analysis framework about economic growth determinants.

Arthur Lewis put forth quite early the framework of economic growth theory, in the preface to *The Theory of Economic Growth* published by him in 1955, he wrote that: "the purpose of writing this book is trying to provide a basis for studying economic growth, to meet the needs of in policy formulation." He also wrote: "this book does not intent to put forth new concepts about economic growth, instead, it tries to provide a relevant basis for the study of economic growth. I want to draw a blueprint instead of setting a theory." In fact, the merit of this works is to draw a blueprint of economic growth by exploring on the growth of per capita output in a country, and many associated factors, such as population, resources, capital, economic system, etc. Lewis believed that the driving factors of economic growth have internal mutual relations, and should be considered as a whole. Of course, "it is true that at a certain place or time, a factor that hinders the growth can be more outstanding than other factors, that is to say, this defect is the most outstanding at this point, or it can be easier to start from this point than starting from other points".[2] However, as the various origins of growth are mutually associated, they must be studied as a whole to avoid misunderstanding.

Later, Simon Kuznets put forth a new theoretical framework on economic growth. In 1973, he published an essay with the title *Modern Economic Growth: Discovery and Thinking* on *The American Economic Review*, providing a new expression of the definition of economic growth: "the economic growth in a country can be defined as the long-term rise of the ability to provide daily increasing economic products to its residents, and such continually increasing ability is founded upon advanced technologies and the required institutional system and the corresponding adjustment of ideology."[3] This expression was innovative on two aspects: first, it stated the sustainability of economic growth, and second, it put forth the analysis framework including put forth including technologies, institutional system and ideology.

In summary, the motive force of economic growth comes from the accumulation and application of production factors and the increase of productive force aspect of production factors. However, for the framework of components, scholars put forth views largely identical but with minor differences, among the literatures on development economics recently published in the United States, Lynn (2009) mainly explained the "source of development" on capital, resources and human capital, and at the same time he discussed industrial technologies, role of the government and foreign trade. The analysis by Weil (2008) involved factor accumulation (practical capital and human capital), productivity (technology, efficiency and opening to the outside world), and basic factors (government, culture, natural environment and resources). Heyne (2008) stated that three factors function in economic development: people, resources and institutional system.

In general, the relevant factors of economic growth are basically the same in both big and small countries. The blueprint described by Arthur Lewis is not out of date yet, and subsequent scholars all further advanced this blueprint. Talking about the distinctions between Large Countries and small countries, mainly, the economy in small countries is concentrated in one or several outstanding industries, such as the processing industry in Singapore, the tourism in Thailand and Maldives, therefore their requirements for natural resources can be unitary or incomplete; but it is different with Large Country economy, its relevant determinants of economic growth are complete, for example, in Large Countries such as the United States, Russia, China, India and Brazil, it is not possible to depend on a single industry to push their economic growth, otherwise, their economic growth could be excessively restricted by foreign factors, which is not conducive to the healthy development of the country. By summing up the relevant research results both at home and abroad, the author proposes the "six-factor structure" for economic growth in Large Countries: natural resources, financial capital, human capital, technological progress, market potential and foreign trade.

5.2 Analysis Framework of Comprehensive Advantages of Large Country

"Comprehensive advantages of Large Country" refer to the comprehensive economic development advantages specific in Large Countries, mainly demonstrated by the economies of scale and division of labor advantages, advantages in regional difference and complementarity, advantages in multiple structure and adaptability, and advantages in independent systems and stability. The framework to describe the comprehensive advantages of Large Country according to the factor structure of economic growth is mainly an organic system formed by the characteristics of natural resources, financial capital, human capital, technological progress, market potentials and foreign trade of Large Countries and their comprehensive advantages. These factors of Large Countries have the general characteristics of large overall scale and difference, heterogeneity and diversity, thereby they produce the effect of scale and adaptability advantage, to promote the rapid, sustained and coordinated development of economy.

5.2.1 Advantages in Natural Resources

Vast territory with abundant natural resources is a basic feature of Large Countries. In terms of national territorial area, Russia ranks at the first place in the world, and China, United States, Brazil and India respectively at the 3rd, 4th, 5th and 7th places in the world. Furthermore, these countries are all Large Countries in inventory of natural resources.

Large gross resources in Large Countries favor the formation of scale economy and pillar industries. According to the principle of comparative advantage, all countries and regions should select industries and products based on the comparative advantage of factors, so that the economy can be competitive. Natural resources are the basic factor, if the selection is based on the comparative advantage of natural resources, Large Countries can have large scale of industries because of their large gross natural resources. For example, the large gross natural resources of China and India for producing textile products have determined the large scale of the textile industry in the two countries, making China the biggest textile producer in the world, and India followed closely, ranking at the third place. The energy output value of Russia accounts for over 30 % of its industrial output value, making the energy industry its pillar sector. Increasing the output of products and forming pillar industries as backed by the advantage of large gross natural resources is a direct embodiment of the natural resource advantage of Large Countries.

The characteristics of Large Countries with rich natural resources can help building an industrial structure with a complete range of categories, and further establishing a complete national economy system. The natural resources in Large Countries are not only large in gross quantity, but also in great varieties. An industrial structure with a complete range of categories can be formed by selecting industries and commodities in the principle of comparative advantage. For example, China ranks at front in the world in mineral reserves of coal, iron, petroleum, antimony, tungsten and tin, and also has resources of agriculture, animal husbandry, forestry, hydropower and aquatic products, which determine that both agriculture and industry are key sectors of national economy, with a complete range of industrial sectors. Brazil is in possession of mineral resources of iron, manganese, alumina, uranium, niobium, lithium, ammonium, graphite and asbestos, and very rich forest and hydropower resources, which determine that it needs to develop various industries. The United States has not only rich mineral resources, but also very rich agricultural resources, so it should not only develop manufacturing industry and high-tech industries, but also develop modern agriculture.

Therefore, Large Countries can establish an industrial structure with a complete range of categories, and have the conditions to focus on developing pillar industries. Such economic development strategy with both diversified industrial structure and pillar industries can well promote the coordinated and steady development of national economy.

5.2.2 *Advantage of Financial Capital*

Large Countries generally have high economic aggregate. In some late-developing Large Countries, although the per capita saving level is not high, the aggregate saving is quite big. Therefore, Large Countries can normally accumulate huge amount of capital, demonstrating the comprehensive advantage of financial capital. For example, China is a developing country, but its rank of economic aggregate in world economy reached the 4th place in 2007. By the end of Apr. 2008, the foreign exchange reserves of China ranked at the first place in the world, exceeding the total foreign exchange reserves in all other countries and regions in northeast Asia, and the foreign exchange reserves of Russia, India and Brazil also respectively ranked at the 3rd, 4th and 7th places in the world. Large scale of capital accumulation and differences in the capital accumulation in different regions are the common economic characteristics of Large Countries. These economic characteristics also demonstrate some economic advantages.

First, the large gross capital accumulation enables Large Country economy to build a self-circulating system, with the ability of self-regulation of capital, to guarantee the stability of the economic system. In an international financial crisis, it usually demonstrates the "Large Country effect" of stable finance. For example, after the serious subprime crisis in the United States, it adopted remedy actions, and launched a bailout program of 700 billion USD, avoiding big turmoil in the society. China could not avoid the impact from this subprime crisis, but it can also take actions with the advantages as a Large Country, and a decision was made to invest RMB 4 trillion yuan to support the development of critical industries and expand domestic demand, to avoid major fluctuation in its economy.

Second, the characteristics of big gross capital and regional differences of Large Countries also enable the country to concentrate capital to support the development of key regions and industries. For example, in early period of reform and opening up in China, to develop the east coastal regions first, the country invested large amount of capital in this region, so that the economy in this region developed rapidly. In late 1990s, to realize coordinated development in the whole country, the state concentrated large amount of capital to support the revitalization of the northeast, the development of the west and the rising of central regions, so that these regions can speed up their development and promote the coordinated development of regional economy. Furthermore, in terms of support to industries, only Large Countries can concentrate large amount of capital to support the development of aerospace industry or other pillar industries, to form scale advantage in industrial development and enhance international competitiveness.

5.2.3 Advantage of Human Capital

Large population is also a basic characteristic of Large Countries. China, India, United States and Brazil rank at the top 4 places in the world in terms of population, and Russia at the 7th place in the world. Of course, developed countries and developing countries have different inventory of human capital, and even within the same Large Country, there are also differences in the quality of human capital between regions, and differences in the types of human capital in different sectors.

The heterogeneity and adaptability of human capital in Large Countries result in comprehensive advantage in human capital in Large Countries. Heterogeneity of human capital exists in Large Countries, either late developing or developed, which means different types of human capital. The economic and social development in Large Countries is usually unbalanced, resulting in different levels of education for labor force, with many high technology professionals and many professionals for applicable technologies, and professionals for manufacturing industry and for modern service industry; they are usually suitable for different production sectors and work posts. In the development of Large Country economy, human capital of different types is suitable to different industrial sectors and making different products. The key point in understanding this issue is that we should not simply classify heterogeneity as superior and interior or advanced and backward, instead, we should see the coupling of heterogeneity with adaptability, that is, the possibility of transforming heterogeneity into adaptability and the inevitability of adaptability demanding for heterogeneity, scientifically analyze this coupling mechanism, and grasp and apply it in practice, so as to realize the organic integration of heterogeneous human capital with the diversified industrial structure and technological structure.

In late-developing Large Countries, the heterogeneity of human capital is especially obvious, and it also demonstrates quite obvious comprehensive advantage in human capital. Some regions have rich high-quality innovation oriented human capital while other regions are abundant in ordinary human capital; some sectors have more high-tech professionals while others have more general-purpose technical professionals; some industries have gathered professionals for physical economy while others have professionals for creative industries. This heterogeneous human capital structure can well suit the coordinated development of economy in late-developing Large Countries. First, the coupling of heterogeneous human capital with industrial structure is conducive to speeding up the optimization and upgrading of industrial structure. Developed regions focus on developing high and new technology industries, to drive up the optimization and upgrading of industrial structure, and gradually transfer the labor intensive industries to underdeveloped regions. Human capital is the motivation to the upgrading of modern industrial structure, the fairly large number of high and new technology professionals in developed regions have become an important driving force in developing high and new technology industries, and the relatively abundant ordinary professionals in underdeveloped regions also provide support to accepting the industries transferred from developed regions. Second, the coupling of heterogeneous human capital with technological level is conducive to fostering the industry competitive advantage in a country. General type human capital can match with comparative advantage industries mainly based on applicable technologies, to meet the demands in economic development and people's life, and high quality innovation type human capital can be gathered in high and new technology and emerging industries, to provide intellectual support to fostering industries with international competitiveness.

5.2.4 Advantage in Technological Progress

The unbalanced development of regional economy in Large Countries and the heterogeneity of human capital result in the multiple structure of technological levels in Large Countries. Large Countries have a complete range of industrial sectors and intact industrial chains, determining the broadness of technological inventory; there are fairly big differences in technological level between industries, with both high and new technologies and applicable technologies; there are also technical gaps within industries, especially in late-developing Large Countries, identical industries in different regions may adopt technologies at different levels. Therefore, in Large Countries, especially late-developing Large Countries, there is usually a multiple technological structure, with high and new technology, traditional technology and handicraft technology coexisting.

Similar to the human resource heterogeneity in Large Countries, the multiple structure of technologies in Large Countries can also be converted into an adaptability. It can adapt to different industries, regions and enterprises, thereby coupling with the development status in different regions, enterprises and industries, to promote the goal of coordinated development. This multiple structure of technologies enables technologies at different levels and of different types to fully play their role in economic development, thereby forming an advantage of technological progress in Large Countries. Also, Large Countries have big economic aggregate and the country and society have powerful ability to invest in strategic technical fields, therefore forces can be concentrated to make innovation in some cutting-edge technologies, to achieve breakthroughs in sophisticated technological fields. For example, in aerospace technology and nuclear energy technology, usually Large Countries such as the United States, Russia and China have advantages, taking the top positions among all countries in the world. This shows that the technologies in Large Countries not only have the advantages needed in the economic development in different regions and sectors, but also the advantage of self-reliance innovation.

Furthermore, late-developing Large Countries have more obvious comprehensive advantages in technological progress. To be specific, the diversification of innovation patterns in late-developing Large Countries is conducive to saving cost in innovation. Practice has proved that the cost of imitative innovation is usually lower than that of self-reliance innovation. The technological innovation in late-developing Large Countries is mainly imitative innovation, therefore they can obtain technologies from developed countries at fairly low cost, gaining the cost advantage in technological innovation. Furthermore, late-developing Large Countries have broad markets, generally more attractive to transnational companies than small countries. The transnational companies in developed countries, via demonstration effect, personnel flow between enterprises and contacts with local enterprises, can promote the digestion and absorption in late-developing Large Countries, giving play to the effect of technology overflow. In the meantime, late-developing Large Countries also have economic strength that favors their self-reliance innovation, therefore they attach importance to the research and development of modern sciences and technologies, to develop the ability of self-reliance development.

This shows that late-developing Large Countries have the "technological imitation—self-reliance innovation oriented" comprehensive advantage, and they can bring into play the respective advantages of importing technology and self-reliance research and development, to fully meet the needs of technological growth in late-developing Large Countries.

5.2.5 Advantages of Market Potentials

Very big domestic consuming demand and market potentials are important characteristics of Large Country economy, and are also the basic factors of CAOLC; consumption pushes economic growth, so in this sense, consumption determine production. The motive force of economic development in Large Countries mainly comes from the demand on domestic market, the United States changing from a big agricultural country to a big industrial country and the rapid sustained development of economy in China all mainly depended on the huge driving force of their domestic markets. Furthermore, with the rising of economic development level, this advantage will become more apparent.

Large Countries have large populations and very large domestic market volume. In 2008, China had a total population of about 1.328 billion, and at the actual level of per capita consumption of 11,242 yuan by urban and rural residents in the whole country, the total expenditure on consumption in that year was 14.92938 trillion yuan, and in 2008, the urban and rural savings deposit in China reached 21.78854, accumulating huge purchasing power.[4] This shows that the huge consuming demand and domestic market in China are not only potential, but also realistic. The United States not only has a large population, the per capita purchasing power is high, and the total consuming demand is big, with powerful domestic investment demand, and the huge domestic market pushed the sustained development of economy. In emerging Large Countries of Russia, India and Brazil, the final consuming demand of residents all ranks at the top positions in the world, and the contribution from domestic demand to economic growth far exceeds that from overseas demand.

The consuming demands in Large Countries are hierarchic. Consumers in Large Countries can be classified into different levels according to different groups, demonstrating differences on regions, nationalities, occupations and income levels. For example, groups of different income levels have demands of different levels on clothing, food, housing and travel; and consumers of different regions, nationalities and occupations also have different requirements. These different consuming groups have different demands, which can be coupled with the hierarchic production levels, product structure and technological levels in Large Countries, thereby promoting the agreement of production structure with consuming structure, to push ahead the flourishing and development of domestic economy.

Moreover, regional differences in Large Countries lead to differences in consumers, giving complementarity to the domestic market. Different regions have different resource endowments, and in a nationwide view, all regions can make products according to the principle of comparative advantage, therefore they can all obtain comparative benefits. This regional production structure can also meet the diversified demands of consumers nationwide, and the advantage of complementation can be formed on domestic market, to build big domestic market in coordinated development.

5.2.6 Advantage in Foreign Trade

Foreign trade is an important determinant in economic growth, and also contains a Large Country effect, and the CAOLC formed by rationally giving play to the Large Country effect. The Large Country effect in foreign trade is mainly demonstrated by the economic scale effect, market monopoly effect and adaptability effect produced by product differences.

The Large Country effect in foreign trade refers to the international market influencing power of a country on a commodity, or the monopolizing power of that country on the market of that commodity. This effect is in two types, the Large Country effect of import and the Large Country effect of export. The former comes from the monopolizing power of buyers, and the latter from the monopolizing power of sellers. A Large Country of import has the feature of large amount of purchase, so its behavior will produce influence on the international market, leading to a positive correlation between the demand of the Large Country and international price; a Large Country of export has the feature of large amount of sales, so its behavior will also produce influence on the international market, leading to a negative correlation between the demand of the Large Country of export and price of the related product on the international market. In the viewpoint of Large Country's interests, the "Large Country effect" produces double results: the influence power of Large Country on the international market can be used, and the supply and demand in the country can be regulated by policies, to improve the trade conditions of the country and obtain benefits to the maximum extent; but it is also possible to cause deterioration of trade conditions because of the dependence on international market, large scale export can cause decrease of prices on international market and large-scale import can cause increase of prices on international market, harming the welfare of the own country. This shows that certain conditions should be created to convert "Large Country effect" into "Large Country advantages". In general, the organization of foreign economic relations and trade should be enhanced, to bring into play the influence on international market and foster the "Large Country advantages".

Large Countries have vast territories and fairly complete industrial systems, and it is necessary to establish a corresponding complete foreign trade system. Especially, the foreign trade of late-developing Large Countries often has a "big and complete" export commodity structure, therefore it can make fairly good use of the natural resources, human resources and economic and technological sources of Large Countries. To a small country, it can comply with the comparative advantage strategy only by establishing a fairly unitary export commodity structure; to a Large Country, even a diversified export commodity structure can also comply with the comparative advantage strategy.

Generally speaking, exporting a single product can gain scale, to obtain scale effect, and diversified export products will restrict the scale, making it difficult to gain international competitiveness. However, Large Countries can be relatively exempted from this law. Large Countries have very big economic aggregate, in some extra-Large Countries, a single region or industrial sector can turn out more than the total output of the whole country or all industries of a small country, therefore, diversified export products structure can still gain scale economy and scale efficiency. Furthermore, in a certain point of view, a relatively diversified export products structure in a Large Country can allow better play of its comparative advantage, so that the advantage of resources diversity of the Large Country can be brought into more effective play and the demands of different countries and groups can be better met, to form a comprehensive advantage and enhance international competitiveness.

5.3 Institutional Innovation and Economic Growth in Large Countries

As mentioned previously, neoclassical economics has already incorporated "institutional system" into the analysis framework of economic growth determinants. The difference of "institutional system" as an economic growth determinant from other factors is that, its role is diffusing and penetrating. The improvement and innovation of institutional system can not only optimize the utilization of natural resources, human capital, financial capital and market potentials, but also promote technological transformation and development of foreign trade, thereby pushing the "comprehensive advantages of Large Country" to transform from potential advantages realistic advantages. Because of this special role with the factor of institutional system, it is analyzed and expounded in a separate section.

5.3.1 Institutional Innovation Pushes Forward Economic and Social Development

In his *An Inquiry into the Nature and Causes of the Wealth of Nations*, Adam Smith built a "simple natural liberty system", or a market pattern of free competition and automatic regulation, which is an institutional system conducive to the growth of national wealth. Marx did an overall survey of division of labor and exchange, production mode and exchange mode, productive force and production relations as well as their contradiction movement. North, the master of neoinstitutional economics, stated that, "in all existing theories describing in detail long-term transitions, Marx's analysis framework is the most convincing". Marx tried to combine technological transition with institutional transition, the interaction of productive force and production relations expounded by him is a pioneering effort to combine the technical limitation and restriction with the limitation of human organizations.[5] In *The Theory of Economic Growth*, Lewis specifically explored on the issue of economic system and growth, i.e. what institutional system, belief or environment is suitable to economic growth, and how the transformation of institutional system influences the evolution of economic growth. North officially incorporated the institutional factor into the analysis framework of economic growth, taking the institutional transition as an endogenous variables of economic growth, and the modern property right theory as the basic theory, he analyzed the relations between institutional transition and economic growth, pointed out that the key to economic growth is institutional factor, and importance should be attached to the promotion or hindering role of policy-based economic factors such as institutional arrangement in economic growth, and he also designed the state model to promote economic growth, and put forth measures to clearly define property rights and reduce transaction cost.

Institutional system is also an important restricting factor in the development of Large Country economy, and institutional innovation plays an important role in the optimization of various factors such as resources, capital, manpower, technology, market as well as foreign trade, and in the formation of "comprehensive advantages of Large Country". Specifically, the general role of institutional factor is mainly demonstrated on three aspects: first, suitable institutional system is the prerequisite for economic growth, as this system can make an environment conducive to economic growth and promote the improvement and optimization of factors associated with economic growth, thereby pushing forward the rapid continual development of economy; second, institutional systems of different types can give rise to different types of economic growth patterns, some promoting the formation of extensive economic growth pattern, and some promoting the formation of intensive economic growth pattern; and third, transformation of institutional system can provide motive force to economic growth and transition, to break the shackle hindering economic growth, or promote the transformation of economic growth pattern. Then, what are the specificities of the roles of institutional factors in Large Countries? They are demonstrated mainly on two aspects in summary: first, institutional transformation in Large Countries is quite difficult and can be started fairly slowly, therefore it is suitable to adopt progressive reform; second, the institutional innovation in Large Countries produces profound influence, with continual pushing force to economic growth, therefore it usually leads to long-term and continual growth of economy.

5.3.2 Reform and Opening up and Rapid Sustained Economic Growth in China

We surely come with such a question when looking at China's Miracle and exploring the cause of rapid sustained economic growth in China for nearly 30 years: China is a Large Country, with the characteristics of vast national territory, large population, rich natural resources and high market potential, and these characteristics are constant. However, why has China, with these same Large Country characteristics, experienced many economic depressions during the recent over 100 years? It goes without saying that one of the important constraint factors is the institutional system, backward and rigid institutional system hindered its economic growth, preventing the transformation of "comprehensive advantages of Large Country" from potential to the real advantage.

The human society develops following a general law, and there is no exception for the development of Large Countries. The historical materialism founded by Marx revealed the general law of the development of human society, and it is the key to understanding the development of Large Countries for us. Marx revealed the law of basic contradiction movement in a given society, that is, the relations of production must be suitable to the property, level and development requirements of productive force, to the relations of production or economic system for realizing the productive force, and function to promote the development of productive force and economic growth; backward relations of production or economic system or those beyond the real productive force would hinder or sabotage the development of productive force and economic growth. With this principle in mind, we can understand fairly well the important role of institutional innovation in Large Countries in pushing forward the economic growth, and answer in a scientific way the question about the tortuosity in the economic development of China. In summary, this is all because institutions or systems not suitable to the real status of productive force constrained the functioning of the economic advantages and potentials as a Large Country in modern China. Also, it is the reform and opening up later that got rid of the constraints by unsuitable economic systems and policies that Chinese economy gained the momentum of rapid sustained growth. Just as Prof. Liu Guoguang said, "We should follow Marx's historical materialist concepts and methods to think about the experience of reform in China, especially, we should conscientiously analyze the experience in 30 years of reform and opening up by applying the principle of basic contradiction movement in a given society."[6]

The reform and opening up launched by the Third Plenum of the 11th Central Committee of the Communist Party of China provided institutional guarantee for giving play to the Large Country advantages of China. The plenum pointed out: "To realize four modernizations, it is required to substantially raise productive force, and it will inevitably require changing in many ways the relations of production and superstructure not suiting the development of productive force, and changing all incommensurate management modes, activity modes and thinking modes, therefore it is an extensive and profound revolution."[7] To overcome the disadvantages of planned economy system and establish the socialist market economy system, China carried out reform of economic system in four phases: in phase I, the economic system reform was started in rural areas by adopting the system of contracted responsibility linking remuneration to output; in phase II, the economic system reform focusing in cities went out all round, and the regulation role of market mechanism on economic operation was gradually strengthened; in phase III, the target mode of socialist market economy was determined, to build up the socialist market economy system on aspects of microscopic economic foundation, market system, administration system of the government, income distribution system and social guarantee system; and in phase IV, the outlook of reform of putting people first and coordinated development was established, to further complete the socialist market economy system. In the meantime, opening to the outside world expanded gradually, to merge the Chinese economy with the international economy; the reform and opening up and institutional transition injected vitality to Chinese economy and pushed forward the accelerated development of Chinese economy. According to the official statistic data, from 1979 to 2007, the actual annual average GDP growth rate of China was close to 10 %; the per capita GDP increased from 173 USD in 1978 to about 2000 USD in 2007, by more than 10 times; the proportion of Chinese economy in the world total increased from less than 1 % in late 1970s to nearly 5 % at the beginning of the 21st century.

As a late-developing Large Country, China realized rapid economic growth for 30 years, and basically at an accelerated rate. It is really worth pondering and summarizing the reasons and experience in it. In author's opinions, the experience is mainly in the following aspects.

First, we adhered to the institutional innovation with fostering market mechanism as the main line, built the socialist market economic system, optimized the economic growth determinants in the Large Country and realized reasonable allocation of various factors such as natural resources, human capital and financial capital. The reform in China is aimed at better releasing the potentials of productive force, so that the various productive force factors constrained under the planned system be activated, and the socialist market economy system be established. From the very beginning, the economic system reform in China was clearly oriented to market, to gradually expand the role of market mechanism, and to rationally allocate various resources with market as the basis. While invigorating the commodity market, we gradually fostered capital market, labor market and land market, reduced the transaction cost of commodities and factors, and raised the production efficiency. In China, a Large Country with rich resources and large population, the rational allocation and high efficiency utilization of production factors with huge scale have pushed ahead the rapid growth of economy.

Second, we adhered to the progressive route of reform, to steadily advance the institutional innovation in a Large Country, avoiding fluctuation and setbacks. China's reform was progressive, i.e. an accumulative marginal evolving transforming mode was adopted in the transition to market economy. The actual process was: reform was started in rural areas, and then extended to cities; it was started in the non-state-owned economy which was more suitable to market economy, and then the focus was placed on the reform of state-owned economy; it was started from invigorating commodity market, and extended to fostering markets of production factors; it was started with the dual-system of planned and non-planned sectors, and then a single market system was realized; and it was started with reform and opening up in special economic zones and coastal cities, and extended to overall reform and opening up in all areas. Just as Prof. Zhang Zhuoyuan pointed out: "the biggest benefit of progress reform is avoiding excessive shock in the society, to push ahead reform and expand opening up while maintaining the society stable, and constrain the adjustment of interest relations resulted from reform within the scope that could be withstand by the society and the public, therefore the relations between reform, development and stability were well handled, to realize balanced transition of economic system."[8] China is a late-developing Large Country, with complicated national conditions, if a "radical" reform mode was adopted, it might lead to social turmoil, increasing the cost of reform; the "progressive" reform mode we adopted could reduce social resistance and realize steady transition.

Third, we combined reform with opening up, so that our Large Country economy could make full use of both domestic and international markets, to drive the economic growth with both domestic demand and international demand. China is a Large Country, with very high market potentials, however, this feature can often lead to a closed economic growth pattern, without using the favorable resources on international market. Foreign trade and international economic cooperation are important factors in economic growth, and also important pushing force to the economic development in Large Countries. With the closed economic mode before reform and opening up, Chinese economy could not participate in the great circulation of international economy, losing many favorable resources and opportunities of development. After adopting the policy of opening to the outside world, actively developing foreign trade, attracting foreign investment and carrying out outbound investment, the import and export volume of China grew at an annual average growth rate of 18 %, the FDI inflow in 2006 was 43 times that of 1985, and the total amount of outbound investment reached 18.7 billion US dollars in 2007. Opening to the outside world not only increased the demand pulling of economy, but also enabled rational allocation of resources in broader range, boosting the prosperity and development of Chinese economy.

Fourth, we adhered to the phased transition in economic growth, so that our Large Country economy could bring into full play both "later-mover advantage" and "early-mover advantage", and realize transformation and upgrading of industrial structure; China has the comprehensive advantages of Large Country as a late-developing Large Country. In general, China is lagged behind developed countries in economic and industrial development, with the "later-mover advantage" to catch up developed countries; also, China has unbalanced economic and technological development level, and some of its industries and technologies have reached the advanced level in the world, with the "early-mover advantage" leading some developed countries. Therefore, the economic transition in China followed a staged route, first, the comparative advantage and later-mover advantage were brought into play, to quickly develop and grow stronger key industries closely linked with national economy and people's livelihood; then the competitive advantage and early-mover advantage were fostered, to develop strategic new industries. In this way, the transition of economic growth was pushed forward by phases, to promote the development of industrial structure from low end to high end.

Fifth, we adhered to the combination of applicable technological with high and new technologies, and adopted both imitative innovation and self-dependent innovation, to drive the development of all factors with technological progress. Science and technology are the primary productive force, and technological progress can lead to high efficiency utilization of natural resources and financial capital, the progress of human capital, improvement of market structure and upgrading of export products structure. For this purpose, China has adhered to the policy of "developing the country through science and education", and implemented the science and technology development policies suiting the national conditions of a Large Country. We not only made full use of applicable technologies to push forward the development of traditional industries, but also actively applied high and new technologies to transform traditional industries, and also fostered emerging industries through technological innovation. In technological innovation, we brought into full play the advantages of a late-developing Large Country and implemented "imitative innovation", to save the cost of technological innovation by using the arbitrage mechanism; meanwhile, we have stuck to the road of self-reliance innovation according to the characteristics of a Large Country, endeavoring to develop core technologies in key industries. In this way, all technological resources are fully utilized, and development of various factors and rapid and sustained growth of economy have been achieved with technological progress.

Notes

1. Adam Smith was an economist first studying the factors of economic growth, with capital accumulation as foundation, he put forth an analysis framework including determinants of natural resources, technological progress, market size and foreign trade. Refer to: Adam Smith: *An Inquiry into the Nature and Causes of the Wealth of Nations*, the Commercial Press 1972 Edition.
2. Arthur Lewis tried to plot a blueprint for economic growth, and put forth an analysis framework including population, resources, capital and institutional system. Refer to: Arthur Lewis: *The Theory of Economic Growth*, the Commercial Press 2002 Edition.
3. Wang Jian: *Transformation and Economic Growth—Research Based on Solow Model*, Fudan University Press 2008 Edition, Page 14.
4. Source: *China Statistical Yearbook 2009*.
5. Hu Dongming: *Economics in A Real World—A Survey of Neoinstitutional Economics*, Contemporary China Publishing House 2002 Edition, Page 7–8.
6. Liu Guoguang: *New Theories of Economics*, Social Sciences Academic Press 2009 Edition, Page 289.
7. Quoted from literature with Shen Baoxiang as chief editor: *Enrichment and Development of Marxism in China since the Third Plenum of the 11th Central Committee of CPC*, Central Party School Press 1987 Edition, Page 66.
8. Zhang Zhuoyuan: *Economic Consideration of China's Experience on Reform and Opening up*, Economic Management Press 2000 Edition, Page 4.

Chapter 6
Strategy: Economic Development Pattern of Large Countries

In the previous chapters, we mentioned the framework of economic growth in Large Countries, and made detailed analysis of the functioning routes of various factors. In this chapter, we will make overall research on the economic development patterns of Large Countries and their characteristics, analyze the suitability of comparative advantage strategy in different phases of Large Country economy development, justify its suitability in the quantity oriented growth phase of Large Country economy and its unsuitability in the quality oriented growth phase of Large Country economy, and put forth the economic development strategic orientation and policy recommendation based on comprehensive advantages of Large Country.

6.1 Characteristics of Economic Models of Large Countries

Scholars both at home and overseas have conducted some researches on the economic development patterns in Large Countries and their characteristics. Simon Kuznets analyzed the effect of country size on the structure of sectors structure, and Hollis B. Chenery analyzed Large Country pattern and small country pattern of development, revealing the characteristics of Large Country economy development patterns; Robert J. Barrow analyzed small country economy pattern, and Gillis analyzed the equilibrium of economic openness of small countries, also revealing on some aspects the characteristics of Large Country economy. Li Daokui analyzed the main feature of Large Country strategy, i.e. the big domestic market and the proportion in world trade, and put forth four measures of Large Country strategy.[1] Jing Xueqing analyzed the main supporting forces for economic growth in Large Countries, and put forth the Large Country economy development pattern with domestic demand as dominant and overseas market demand as auxiliary.[2] Yu Weihua et la analyzed the possible problems in late developing Large Countries

Y. Ouyang, *The Development of BRIC and the Large Country Advantage*,
DOI 10.1007/978-981-10-0633-3_6

adopting the comparative advantage development strategy, and put forth the countermeasures for utilizing the advantages of Large Countries, and overcoming Large Country disadvantages.[3] Lu Ming analyzed the characteristics in Large Country governance, revealing the world significance of China's road of Large Country development.[4] Yang Rudai and Yao Yang analyzed the Large Country characteristics of China, and put forth the "limited catch-up" strategy.[5] It can be seen that the Large Country economy development strategy and patterns have become a hot issue in the researches of domestic academic circle.

After making reference to the research results of both domestic and overseas scholars, the author came to the view that, given the special features of Large Country economy, the characteristics of Large Country economy development patterns are mainly demonstrated on the following aspects:

First, given the scale of economic development in Large Countries, the strategy of overall advancing with breakthroughs in key areas should be adopted for industries. Large Countries have vast territories, with large space for economic activities and industrial distribution; Large Countries have rich resources, able to provide sufficient production factors for different industrial sectors; and Large Countries have broad markets, able to form consuming demands to drive up different industrial sectors. All these special features usually enable Large Countries to bring into play their comparative advantage and economies of scale advantage, to establish relatively complete and independent national economy systems. In the meantime, these different features also requires Large Countries establishing complete economic and industrial systems, to maintain steady and coordinated development of economy. In a Large Country, the basic needs of economic development and people's living cannot be met without some important industrial sectors, especially those linked with national economy and people's livelihood, such as agriculture or energy, and such an economic system would be defective, in a very unfavorable situation to the development domestic economy, and could be easily controlled by others in the international economic competition. Therefore, Large Countries should adopt the strategy of overall advancement in industries, maintaining the all-round development of main industrial sectors and the coordinated stability of domestic economy; in the overall advancement, the comparative advantage should also be brought into full play, to realize breakthroughs in key sectors, and great efforts should be made to support industrial sectors that can gain international competitiveness. In fact, all Large Countries in the world adopt the strategy of "overall advancement with breakthroughs in key sectors". For example, the economic system of the United States is a typical complete system, not only with a complete range of industrial sectors of industry, agriculture and service industry, as well as manufacturing, energy, chemical, aerospace and biology, most of sectors are also quite powerful. Russia has a solid industrial foundation with complete range of sectors, with heavy industry and energy industry especially outstanding, however, its light industry is not sufficiently developed, and affected the healthy development of economy to a certain extent. After the founding of New China, a typical overall advancing economic development strategy was followed, and independent industrial systems and national economy system covering a complete range of categories were quickly established. After the WWII, Brazil built up a fairly complete industrial system by implementing the "import substituting" strategy. After the war, India also achieved fairly rapid economic development, and gradually established an industrial system including iron and steel, smelting, machine building, chemical, energy as well as agriculture and service industry.

Second, given the differences of economic development in Large Countries, the strategy of coordinated development should be adopted for regions. Large Countries have large spaces, and all regions have differences or unbalance in geographic advantages, natural resources, human capital as well as capital conditions. This special feature of Large Countries surely demands regional coordinated development strategy, to realize coordinated development of regional economy with correct policy orientation, so as to maintain the economic prosperity and social stability in the country. The economic development in China has obvious regional differences, as demonstrated by the unbalanced economic development in the east region, central region and west region. The Party and state have always adhered to the policy of overall planning and all-round development to maintain the prosperity and stability of economy in the whole country. Especially, during the reform and opening up, the "gradient transfer" strategy was adopted, first, with the preferential policies of the state, the east coastal region was developed first, as a demonstration to the economic development in the whole country, then, strategies of "developing the west", "rise of Central China" and "revitalizing the northeast" were implemented, to realize the coordinated and steady development of national economy. In fact, the United States was a country with unbalanced development of regional economy, and according to the data of 1994, Alaska had the highest per capita GDP, while Mississippi was the lowest, with a gap of 3.2 times between the two states. The US Government implemented policies and measures to develop underdeveloped regions, so that the economic development in different regions has become fairly balanced. At present, countries with fairly high differences between domestic regions are India and Brazil. According to official statistics, the differences of per capita GDP in 16 regions of India has expanded since the 1990s, the per capita GDP in rich regions is about 5.6 times that of poor ones, the development level of the east coastal region of standard is equivalent to that of the south, but the central and western regions are dominated by agriculture, with fairly low productive forces, still in a backward state. In general, the space characteristics of Large Countries have determined their unbalanced regional economic development, but the prosperity and stability of country require coordinated development of regional economy, therefore, Large Country economy strategy should be aimed at coordinated development of regions, to achieve relative balance for regional economies in absolute unbalance.

Third, given the endogeneity of economic development in Large Countries, the strategy of internal and external circulation should be adopted in opening up. Barrow pointed out when analyzing economy in small countries that: "if a small country economy is in very small scale with respect to world economy, then the accumulation of assets and capital inventory of that country has minor influence on the world interest rate r(t) route. Therefore, to a small country, the r(t) route can be taken as exogenous. Relatively speaking, Large Country economy can be taken as endogenous."[6] The characteristics of Large Countries of abundant natural resources, human capital and capital as well as high market potentials have determined that it is an endogenous economic pattern, i.e. an economic pattern dominated by domestic demand with overseas market demand as auxiliary; some scholars took three Large Countries of China, the United States and Russia as objects of study, to calculate the gray correlation coefficient and gray correlation of the GDP growth rate in these countries with the contribution from total investment, contribution from total consumption, contribution from net export by applying the gray correlation analysis principle, to reveal the internal links of these four indicators, as shown in Table 6.1.

It can be seen from the above gray correlation that, the correlation of contribution from total investment and contribution from total consumption with GDP growth rate in the three Large Countries is in general higher than that of contribution from net export with GDP growth rate, which shows that the contribution from domestic demand to economic growth in Large Countries is generally higher than that of overseas demand to economic growth, and the Large Country economy is mainly an endogenous economy.

As shown in Table 6.2, the economic growth in Large Countries are mainly pushed by their domestic market demands, therefore, the economic development in Large Countries should be focused on domestic consuming demands, and mainly depend on the expansion of domestic demand scale. In the meantime, with the trend of economic globalization, Large Countries must not seclude them from the outside world, instead, they should take an active part in international economic exchange and make use of both domestic and international resources and markets to the maximum extent. In short, the "internal and external circulation" strategy should be adopted, that is, establishing a mechanism on an open economic system, when the international economic environment is good, the external circulation can drive the internal circulation, to push forward the development of the economic system; when the international environment is not good while the domestic economic environment is good, it can depend on domestic demand and perform self-regulation, to

Table 6.1 Gray correlation coefficient and gray correlation of GDP growth rate in different countries with contribution from total investment, contribution from total consumption, contribution from net export

Country indicator	China	United States	Russia
GDP growth rate	1	1	1
Contribution from total investment	0.590	0.657	0.649
Contribution from total consumption	0.563	0.777	0.653
Contribution from net export	0.378	0.738	0.363

Note The calculation was based on relevant data from *International Statistical Yearbooks* (2001, 2003, 2004 and 2005), Quoted from *Theory of Internal and External Economic Circulation and Large Country Economy Development Strategy* by Zeng Jianqiu and Ding Ke, *Journal of Beijing University of Posts and Telecommunications* (*Social Science Edition*) 2007 Vol. 9 No. 3

Table 6.2 Contribution rate of overseas demand and domestic demand to economic growth in China (*Unit %*)

Year	Proportion of demands in GDP		Contribution rate of growth of various demands to GDP growth		
	Domestic demand	Overseas demand	GDP growth	Domestic demand	Overseas demand
1991	97.1	2.9	9.2	96.4	3.6
1992	98.9	1.1	14.2	107.5	−7.5
1993	102.0	−2.0	13.5	111.1	−11.1
1994	98.7	1.3	12.6	89.6	10.4
1995	98.3	1.7	10.5	97.0	3.0
1996	97.9	2.1	9.6	95.4	4.6
1997	96.4	3.6	8.8	80.8	19.2

Source Han Wenxiu and Fan Jianping: *Expanding Domestic Demand*: *New Deal of China*, China Materials Press 1999

push forward the development of the economic system; when both international and domestic economic environments are good, it can depend on the favorable circulation of both internal and external systems, to push forward the development of the economic system.

Fourth, given the autonomy of economic development in Large Countries, the strategy of active progress should be adopted. The economic endogeneity of Large Countries has determined its autonomy, and the internal circulation system of Large Countries provides an important prerequisite for its autonomy. The economic and market scale of Large Countries enable them to produce greater influence on the development of world economy, relatively speaking, small country economy can only passively accept the impact from world economy, but Large Countries can take initiative to influence the world economy. The US economy ranks at the first place in the world in scale, as the locomotive to drive the development of global economy, and its economic movement can produce far-reaching influence on world economy, for example, the economic slump in the United States in 2000 triggered the economic depression in the EU, Japan and Canada, and the subprime crisis in the United States in 2007 triggered a global financial crisis. In these years, the international influence of emerging Large Countries China, Russia, India and Brazil increased, and they made important contributions to the economic growth in the world. Especially, in the 2008 financial crisis, governments of all countries implemented programs to promote steady development of economy, making active efforts to bail out the global economy. The autonomy of economy development of Large Countries and the important influence on international economy have determined their important responsibilities to lead the economic development in the world, to well implement their responsibilities as Large Countries, they should adopt an active progressing strategy. First, they should be active in driving the economic development. Large Countries should bring into play their scale advantages and special influences, take an active part in world economic competition, promote the establishment of a fair, impartial, tolerating and well-organized new international economic order, and drive the sustainable development of world economy with advanced development concept. Second, they should endeavor to push forward technological innovation. Large Countries have powerful economic strength, and should concentrate forces to solve technical difficulties and realize innovation and breakthroughs in industrial technologies, to drive the development of global economy with technological progress.

6.2 Suitability of Comparative Advantage Strategy and the Staged Economic Growth of Large Countries

Since the start of reform and opening up, the Chinese economy has mainly depended on factor accumulation to push its growth, and the contribution from technological progress was low, in the factor input quantity driven economic growth stage.[7] However, this extensive growth pattern has produced important impact of resources and energy shortage and environmental deterioration, in the meantime, this extensive economy of China has also resulted in deterioration of trade conditions: many industries and even critical sectors are lacking in competitive advantage, foreign businesses have controlled many industries of China, and China is at the low end of value chain of international division of labor, with low economic gains and difficulties to boost domestic demand, this situation has made it inevitable for China to change its economic growth pattern. Therefore, the *Eleventh Five-year Plan Outline for National Economic and Social Development of the People's Republic of China* clearly stated that "economic growth shall change towards an intensive growth pattern", and that "we should take the new road to industrialization". Then, what development strategy orientation should China, as a developing Large Country, take in the transition of economic growth pattern? During reform and opening up, China adopted a development strategy oriented by comparative advantage, and this strategy greatly stimulated the rapid growth of Chinese economy, and raised the trade share and position of China on the world market. However, is the comparative advantage strategy still suitable to the transition of economic growth pattern and new industrialization construction of China, and how big is the space of this strategy in the transition of economic growth pattern? This section will discuss on these questions.

6.2.1 Comparative Advantage Strategy Theory and the Questioning on It

The economic development strategy of a country can make it flourishing and also make it declining. Since the industrial revolution in the 18th century, the gap between countries in economic development has widened up, therefore, what development strategy and policies underdeveloped countries should take to realize industrialization and catch up and surpass developed countries became a pressing subject to the political leaders and intellectuals of these countries. Many economic growth theories and literatures on development economics paid high attention to this issue, and various propositions and doctrines have also been produced. Among many ideological views, Justin Yifu Lin et al. systematically put forth the highly influential "comparative advantage strategy theory",[8] stating that "development strategy is an important factor to determine if convergence can take place in the country, most under-developed countries failed to successfully narrow the gap of development with developed countries mainly because their governments adopted unsuitable development strategies".[9] This is because: "the per capita revenue of a country is a function of its technology and industries", "the root cause of developed countries being rich lies in their advantages in industries and technologies"[10]. "The fundamental difference between a backward country and a developed country lies in the difference in factor endowment structure." "If the government of a developing country chose to develop in priority the industrial/technical structure not in compliance with the comparative advantage determined by the factor endowment of the economy, on a competing market, enterprises in sectors developing in priority would lack in viability. …As a result, the development performance of this economy will be quite poor, and no convergence will take place. Only when the government of a developing country takes comparative advantage as the basic criterion for industrial development, this economy can have a well-operating market, and it can easily import technologies from developed countries, maintain a high capital accumulation rate, achieve rapid upgrading of factor endowment structure and realize convergence".[11] Therefore, "governments of under-developed countries should aim at promoting upgrading of factor endowment structure, instead of upgrading of structure, because once the factor endowment structure has been upgraded, the motivation for profit and competition pressure will drive enterprises to spontaneously upgrade their technological and industrial structures".[12]

It can be seen from the above analysis that, the "comparative advantage strategy theory" has clearly built the analysis paradigm of "factor endowment—viability—selection of technological and economic structure—economic growth (convergence)", that is, economic growth depends on technological and economic structure, which in turn is endogenous from factor structure (capital to labor ratio). Failure to follow the comparative advantage of factor structure will result in no viability with enterprises, and poor performance in economic development. This theory is based on some experience and lessons in developing countries implementing development patterns and strategic transition in recent decades, and by the summary of massive historical evidences, it has provided a "nearly complete" theoretical explanation to the practices of industrialization process in China and in all developing countries.

However, since the founding of the comparative advantage strategy theory, the suitability of this theory to the economic development in China has been questioned on many aspects, mainly on the following: (1) Can it achieve fairly high performance of economic development? The continual deterioration of price trade conditions of China since the 1980s has given rise to questions about the effectiveness of the comparative advantage strategy.[13] According to comparative advantage strategy, taking labor intensive industries as the current dominating industries, under the modern trade pattern, cannot obtain high performance of economic development, therefore there will be no much economic surplus left to realize upgrading of factor endowment structure.[14] (2) Can it help raise the competitiveness? Guomin and Yongqin (2003) pointed out that the comparative advantage of resource endowment is not the sufficient and necessary conditions of the industrial competitiveness of a country and abilities of enterprises, lacking of technological advantage has made it even more difficult to continue the comparative advantage of the labor intensive industries in China. Yinxing (1997) also believed that pure comparative advantage will not surely become competitive advantage. (3) Is it conducive to upgrading of industrial structure? Kesha (2004) pointed out that a fundamental question of comparative advantage strategy theory is "whether it can realize the transition of dominating industries and whether it is conducive to the transformation of comparative advantage". Although Lin (2002) pointed out that "the Four Asian Tigers" smoothly realized upgrading of industrial structure by improving the factor endowment structure, China has its special feature as a developing Large Country, as it is lacking in capital but is quite abundant in labor force. And more important, due to the existence of the dual economic structure, with the industrial restructuring and completion of market economic system, massive labor force moved out of rural areas, so that the labor intensive comparative advantage industries in China have almost unlimited supply of labor force, which makes it very difficult to improve the factor endowment in China in a short period of time, resulting in the rigidity of industrial structure upgrading in China. Furthermore, as far as trade structure is concerned, Xibao and Hanchang (2004) pointed out that labor intensive comparative advantage industries cannot automatically and spontaneously transit to capital intensive and technology intensive industries.[15] A late-developing Large Country like China, if participating in international division of labor in the principle of comparative advantage, will remain forever an export country of primary products and labor-intensive products. But labor intensive industries cannot drive the development of the whole economy, therefore it cannot realize sustained and rapid development of economy by only relying upon a limited number of labor-intensive products, as "the Four Asian Tigers" did.

Therefore in our opinion, the "comparative advantage strategy theory" has well explained the economic development history of China, by following the comparative advantage theory, China has sustained rapid growth for nearly 30 years, and obtained considerable economic surplus. However, China is facing the important task of transforming the economic growth pattern, and given the questioning on the comparative advantage strategy theory, it is necessary to make further systematic exploration on the suitability of this theory, to make clear the theoretical logic sources of the many questions on this theory, and explore the development strategy orientation in the transition of economic growth pattern in China. Following the basic principle put forth by Prof. Justin Yifu Lin that "the criticism of a theory is directed either to its consistency of internal logics or to the consistency between its logical deduction and empirical fact", the author now makes systematic analysis on the suitability of the comparative advantage strategy theory in different stages of economic growth.

6.2.2 *Law of Staged Economic Growth and Structural Transfer of Leading Industries in Large Countries*

Due to the complicated economic structure and factors, Large Countries cannot realize leap-forward development of economy with the effect of external factors, as with small countries, instead, they can only advance progressively by stages, and the economic growth pattern will go from the labor force input driven stage, to capital input driven stage, and then further to knowledge technology driven stage, this is the law of staged economic growth in Large Countries.

In either the history of world economy growth or the history of development of economic growth theories, the sources or routes of economic growth are only on two aspects: one is the quantity of production factors input (including natural resources, labor force and material capital); and the second is the efficiency of utilizing the production factors, or the so-called productivity (including the improvement of general efficiency and the upgrading of factor quality). The different compositions of these two growth sources, that is, economic growth depending mainly on increasing factor input (extensive and quantity-oriented) or on raising factor productivity (intensive and quality oriented) actually reflects an essential difference of the growth patterns, normally it is believed that it is basically an extensive (quantity-oriented) economic growth if the contribution from total factor productivity growth rate to economic growth rate is less than 50 %; and it is an intensive (quality oriented) economic growth if this contribution is 50 % or more.[16] The history of world economic growth shows that the law of economic growth pattern and leading industry structure transformation is as shown in Fig. 6.1.

(1) Transformation from agricultural economy to labor input driven economic growth and its leading industrial structure. Since the first industrial revolution, the economic development history of developed countries has demonstrated the characteristics of staged economic growth pattern and transformation of leading industrial structure. Before the first industrial revolution, the agricultural economy dominated, and mainly depended on natural resources such as land to promote economic growth. After the first industrial revolution, during about 1820–1860, the labor-intensive light industry dominated by textile industry became the pillar industry in the economy. This period was referred as labor force input driven economic growth stage.[17] During 1760–1860, with the continual rising of agricultural labor productivity, agriculture was able to

要素投入数量驱动型(数量型)
要素使用效率驱动型(数量型)
资源驱动型 劳动力驱动型
资本驱动型 管理、知识、技术驱动型
农业经济主导　劳动密集型产业主导
资本密集型产业主导
知识、技术密集型产业主导
生产率提高

Factor input quantity driven (quantity-oriented)
Production factor application efficiency driven (quantity-oriented)
Resources driven　Labor force driven
Capital driven　Management, knowledge and technology driven
Agricultural economy leading　Labor intensive industry leading
Capital-intensive industry leading
Knowledge, technology-intensive industry leading
Increase of productivity

Fig. 6.1 Transformation of economic growth pattern and leading industrial structure

provide raw materials for the development of light industry and labor force emancipated from rural areas, thereby promoting the development of light industry. Therefore in the initial period of shifting from agricultural economy to industrial economy, the prerequisite for a country to develop manufacturing industry should be increasing the labor productivity of agriculture.

(2) Transformation from labor input driven to capital input driven economic growth stage and its leading industrial structure. After the first industrial revolution, the rising of productive force promoted the development of light industry, and the development of light industry in turn accumulated massive material capital, paving the way for the development of capital-intensive heavy and chemical industries. In the meantime, the development of agriculture and light industry significantly increased per capita GDP, and the increased income generated demand from people for heavy industrial products with fairly big demand elasticity. Therefore, pushed by the second industrial revolution, the heavy and chemical industries dominated by iron and steel, machine building, coal mining, power, ship-building, automobile and petrochemical developed quickly, becoming leading industries. For example, the proportion of heavy and chemical industries in the whole industry reached 52.9 % in 1939 in the United States, 56.3 % in 1960 in Japan, and 60.8 % in 1950 in Federal Germany, in terms of growth pattern, this period was still in the factor input quantity driven economic growth stage, and capital accumulation was the main route of economic growth. In industrialized countries of the United States, United Kingdom, Germany and France, the average contribution of scientific and technological factors to economic growth was only about 30 %, for

example, during 1840–1900, the productivity contributed only 17 % to the growth of net national product, while factor input accounted for 82 %.[18]

(3) Transformation from capital input driven to knowledge and technology driven economic growth stage and its leading industrial structure. The rapid development of capital-intensive heavy and chemical industries driven by the second industrial revolution produced effects on two aspects: first, the world resources and energy consumption increased sharply, causing shortage of resources and energy and deterioration of environment; second, with the further increase of capital accumulation, the law of diminishing marginal return from capital input became prominent, demanding seeking for new sustained growth points in economy. Driven by these two effects, developed countries started developing technology and knowledge-intensive industries energetically. Therefore, in about 1950s, the modern scientific and technological revolution with micro electronic technology, new energy technology and new material technology as the core became widespread, and this, together with scientific management marked by "expert system", greatly promoted the transformation from capital input driven to knowledge and technology driven economic growth stage in developed countries. For example, during 1950–1960, the GDP annual average growth rate in the United States was 3.3 %, to which the contribution from factor input was only 47 %, while that from total factor productivity was 53 %, indicating the realization of growth pattern transformation. After that, the Federal Germany in the 1960s and the UK, France and Japan in the 1970s realized in succession the transformation from factor input quantity driven to production factor application efficiency driven in their growth patterns.

In general, the history of world economic development shows that, in countries with fairly big sizes, the economic development all followed such a law: the economic growth pattern transformed from "labor input driven" to "capital input driven", and then further to "technology and knowledge driven", and along with the transformation of economic growth pattern, leading industries transformed from labor-intensive to capital-intensive and further to technology-intensive industries.

6.2.3 *Suitability of Comparative Advantage Strategy in Quantity-Oriented Economic Growth Stage*

In the transformation from labor force input driven economic growth stage to capital input driven economic growth stage, i.e. in the stage of leading industry structure transforming from labor-intensive to capital-intensive, the comparative advantage development strategy theory is suitable. The root cause is that capital accumulation is the critical factor in realizing this transformation. Both the internal logic system of "factor endowment—viability—technology and economic structure selection—economic growth" and the consistency of this theory with the realistic experience indicate the suitability of comparative advantage strategy theory.

1. The decisive factor of enterprise viability is the factor cost

In the factor input quantity driven economic growth stage, capital is the most important production factor. The factor endowment structure (relative abundance of capital) determines the capital cost, if the capital is relatively abundant, the capital cost is low, and vice versa. At this time, the viability of an enterprise depends on whether the types of products made comply with the resource endowment of the whole society, i.e. with the comparative advantage of economy. When capital is relatively scarce, "enterprises in under-developed countries would enter labor intensive industries, to select relatively labor-intensive technologies in production", on the contrary, "the capital-intensive industries that the government wants to develop in priority are not in line with the comparative advantage of this economy, and enterprises in these industries have no viability. Then, to support these enterprises without viability, the government would take a series of distorted measures on international trade, financial sectors and labor market. Although it is possible to establish capital-intensive industries in a developing economy by distortion, it will result in distorted resources allocation, rampant rent-seeking activities and unstable macroeconomy, and low efficiency in economy, as a result, the target of convergence cannot be realized".[19]

2. Industrial structure is endogenous in factor endowment structure

The transformation of industrial structure from labor-intensive to capital-intensive depends on the relative abundance of capital, "if under-developed countries develop industries by following the comparative advantage, there would be maximum possible economic surplus and highest tendency of saving, therefore the maximum possibility of factor endowment structure upgrading." "Once the factor endowment structure has been upgraded, the profit motivation and competition pressure will drive the enterprises to spontaneously upgrade their technologies and industrial structure. Factor endowment structure upgrading means capital accumulation grows faster than the growth of labor and natural resources, and this is the case for both material capital and human capital." In the meantime, "when development follows the comparative advantage, the production of factories is in a state of maximization of profit, with the most economic surplus for accumulation. In the meantime, as the factor prices correctly reflect the scarcity of factors, in the situation of relatively scarce capital, capitals have relatively higher prices, and this greatly motivates the saving desires of residents, therefore both accumulation rate and per capita accumulation level can meet the needs in sustained economic growth." Therefore, "governments of under-developed countries should aim at promoting structure upgrading of factor endowment, instead of upgrading industrial and technological structures".[20]

3. Consistency of comparative advantage strategy theory and empirical facts

First, the economic growth history of developed countries shows that, the transformation process from an agricultural economy to a labor force input driven economy, and further to a capital input driven economic growth stage is fairly in line with the inference of comparative advantage strategy theory, that is, with the increase of capital abundance, its leading industry will change from agriculture to labor-intensive light industry and further to capital-intensive heavy and chemical industries. Second, according to the catch-up experience of the former Soviet Union and after the founding of New China, the development of capital-intensive heavy and chemical industries under the conditions of lacking in capital really achieved no good results as this practice went against the comparative advantage strategy theory. Furthermore, as demonstrated by the emerging the Four Asian Tigers, their industrial structure was also transformed from labor intensive industries to capital-intensive industries with the improvement of factor endowment structure.

In general, both the internal logic system of "factor endowment—viability—technology and economic structure—economic growth" and the consistency of this theory with the realistic experience indicate that the comparative advantage strategy theory is suitable in the factor input quantity driven economic growth stage[21]. Therefore in the selection of development strategy, just as Lin and Mingxing (2003) said, "to realize the goal of transforming from an agricultural country to an industrial country, developing countries should follow the development strategy of comparative advantage of their own countries".

6.2.4 Unsuitability of Comparative Advantage Strategy in Quality Oriented Economic Growth Stage

In the society today, the economic growth pattern is undergoing fundamental changes, i.e. from factor input quantity driven (quantity-oriented) to production factor application efficiency driven (quality oriented), a large number of developing countries having merged into the global economy are possibly still in the quantity-oriented economic growth stage, but the developed countries already in the quality oriented economic growth are competing with them, resulting in change of the environment for analyzing the comparative advantage theory. In the quality oriented economic growth stage, both the internal logic system of "factor endowment—viability—technology and economic structure selection—economic growth" of comparative advantage strategy theory and the consistency of this theory with the realistic experience indicate the unsuitability of the comparative advantage strategy theory in this economic growth stage.

1. The determinant of enterprise viability: factor cost or application efficiency of factors

The enterprise viability, that is, the ability to earn profit at a level acceptable on the market without protection or allowance from the government, is a critical variable to link factor endowment structure with industrial structure and economic growth in the comparative advantage strategy theory, in which capital cost (interest rate) is a critical factor determining the enterprise viability because mainly the factor of capital is taken into consideration. However, in the quality oriented economic growth stage, as the contribution from application efficiency of production factors (or TFP) to output has exceeded that from capital, does the determinant of enterprise viability remain as the factor cost?

To answer this question, let's start from the case of India economy with which we are fairly familiar. In resource endowment structure, the saving rate is about 24 % in India, as compared with 40 % in China, and the FDI inflow to India every year is only 10 % that of China (Diana Farrell et al. 2004), so we can obtain the basic conclusion that "China has more abundant capital factor than India". According to the comparative advantage strategy theory, India's industrial structure should be lower than that of China, and relative to China, the economic growth in India should be in an extensive pattern. However, as compared with the extensive economic growth pattern (factor input quantity driven) of China, India is generally recognized as an intensive (production factor application efficiency driven) economic growth pattern with high input-output efficiency, while India is known as the "world office", China is known as the "world factory". Over a long time, China has been following a typical "East Asian Model", i.e., recirculating the export revenue to the investment on fixed assets, to drive economy with investment. However, the growth pattern of India is more similar to the Anglo-Saxon model, instead of East Asian Model.

According to the comparative advantage strategy theory, China has more abundant capital factor than India, so the technology- and knowledge-intensive enterprises in China should exceed those in India in both competence and number. But actually, India has fostered more successfully a large number of private enterprises and companies with international competitiveness, and most of them are based on the edge-cutting information technologies, such as the famous Inosys and Wipro on software, and the famous Ranbaxy and Dr. Reddy's Labs in pharmaceutical and biological technologies. According to the *World Competitiveness Annual Report of 2003 of International Development Research Institute in* Lausanne of Switzerland, although China's ranking in terms of comprehensive competitiveness is ahead of India, the ranking of business efficiency of India is 5 places ahead of China, mainly because the microscopic foundation for the economic growth in India is mainly some private enterprises full of vitality and with fair competitiveness in the world.

Why can India, with less capital factor than China, foster technology and knowledge intensive enterprises more successfully than China? It is in essence that the enterprise viability of technology and knowledge intensive industries is not solely determined by factor cost. In fact, in the trend that the world economy is shifting to the production factor application efficiency driven economic growth stage, the determinant of enterprise viability has also changed, from factor cost to application efficiency of production factors. In the production factor application efficiency driven economic growth stage, the TFP contribution in macroscopic gross quantity can exceed that from factors only when the contribution from application efficiency of production factors to the sales income of an enterprise exceeds that from factor input quantity, therefore in this stage, the determinant of enterprise viability is productivity instead of factor cost. The endogenous growth theory provides a fairly detailed explanation of the application efficiency of production factors: this theory states that the application efficiency of production factors, or productivity can be determined by two factors: technology[22] and efficiency. Technologies can be obtained by research and development and international technological diffusion (Barro 1997), while efficiency is a general concept, i.e. the effectiveness obtained by combining production factors with technologies, and it is used to describe and interpret difference in productivity, instead of anything in technological difference. Its influence factors are also all-inclusive: the production organization forms, non-production activities (such as economic rent-seeking), idle resources (the production factors are not fully utilized) and management in enterprises.

Of course, investment is also required to raise the application efficiency of factors, for example, large amount of investment is needed in technological improvement. Vindicators of comparative advantage strategy may respond this way: large amount of investment is inevitably needed to create advanced production factors (such as technologies). The sources of investment can only be the economic surplus created in the past production activities of enterprises and the whole economy. Justin Yifu Lin pointed out that, only when the production activities are organized according to comparative advantage, the enterprises and the whole economy can create economic surplus to the maximum extent, even on a single factor that influences the application efficiency of production factors—the technical aspect, a leading technical position is sufficient to make up the cost of capital. Because according to the basic principle of microeconomics, market of new technology products has the monopoly feature, and the high profit obtained from monopolized pricing can make up the capital loan cost—the interest rate. In addition, in the production of technology, the determinant of new technologies is high quality human capital, instead of material capital. Therefore salaries for research and development personnel account for the most part of R&D cost, as shown by American Science and Engineering Indicators, in the R&D in colleges and universities, the expenditure on R&D equipment only accounts for 5–7 % of the total R&D cost.[23] High quality human capital is fostered by the public education of the government, its quantity and quality is an exogenous variable to the decision-making in an enterprise, therefore the government can increase the quantity and quality of high-quality human capital even under a condition of relatively lacking in capital, to reduce the R&D input of enterprises[24]. The fact that the high quality elite education in India has fostered large number of software professionals is a very good example. That is why India has so many high-tech enterprises with international competitiveness under the conditions of not quite abundant capital—the application efficiency of production factors really determines the viability of enterprises.

About the explanation of enterprise viability, or the source of enterprise competitive advantages, there are many theoretical views, and not all of them regard the factor cost advantages as a determinant. For example, the resource school believes that the sources of enterprises' competitive advantage lies in the strategic resources controlled by them, such as the technological knowledge, highly skilled employees, brands, trade links and capital of enterprises. The dynamic capability school believes that the most precious asset of enterprises is the capability based on organizational knowledge, and how to develop, maintain and enhance the organizational capability has a critical effect on enterprises in winning competitive advantage.

Now let's look at the questioning upon comparative advantage strategy theory, the logic source of questioning comparative advantage strategy theory based on deterioration of trade conditions and lacking in competitiveness also lies in the different understandings of the determinants of enterprise viability. This is because, in the trend of international economy moving towards "quality oriented" economic growth stage, the determinant of enterprise viability is not the comparative advantage of factor endowment, instead, it is the increased application efficiency of production factors resulted from leading knowledge and technologies, therefore, exporting purely on the basis of comparative advantage will naturally result in poor trade conditions, and naturally there will be no competitiveness. Therefore, naturally, comparative advantage strategy theory received lots of questioning such as the "deterioration of trade conditions and effectiveness of comparative advantage strategy[25]"; "high economic development performance cannot be obtained under the modern trade pattern by taking labor intensive industries as current leading industries according to comparative advantage strategy"[26]; "pure comparative advantage will not necessarily become competitive advantage"[27]; "the comparative advantage of resource endowment is not the sufficient necessary condition of the industrial competitiveness of a country and the viability of enterprises themselves, and developing countries should pursue for strengthened innovation and breakthroughs in many critical technologies … instead of deliberately pursuing for the so-called comparative advantage strategy"[28].

2. Whether technological and industrial structures are surely endogenous in factor structure

The comparative advantage strategy theory following the analysis paradigm of "factor endowment—viability—technology and economic structure selection—economic growth" believes that technological and industrial structures are surely endogenous in factor structure with the reason that: "once the factor endowment structure has upgraded, the profit motivation and competition pressure will drive enterprises to spontaneously upgrade their technological and industrial structures"[29], in the factor input quantity driven economic growth stage, it is true that technological and industrial structures are endogenous in factor structure, because capital is a decisive factor in this stage. However, in the production factor application efficiency driven economic growth stage, are the technological and industrial structures surely endogenous in the factor structure? That is, is it true that the higher factor abundance, the higher technological level and industrial structure?

First, in the relations between technological structure[30] and factor endowment, the comparative advantage strategy theory believes that technological structure is endogenous in factor endowment on the ground of "suitable technology" theory of Basu and Weil (1998). For example, "for enterprises in under-developed countries, the industries and technologies to be upgraded are new, and need to be transferred from developed countries. The cost of learning is lower when following the comparative advantage strategy than going against the comparative advantage strategy",[31] so, with the upgrading of factor endowment structure, it will be possible to import higher technological structures suitable to the factor endowment structure, therefore, developing countries "should aim at promoting the structural upgrading of factor endowment, instead of upgrading the industrial and technological structures". But there are two fundamental questions in this logic: the first question, if technologies can be imported, why the enterprises in so many developing countries did not import technologies "suitable to their factor endowment structure" as driven by the "profit motivation and competition pressure", to realize technological and economic convergence? The fact is that, "the developing countries in large numbers have not realized economic convergence by importing technologies, and the gap from developed countries in per capita GDP has expanded instead". Is that only because these countries have all adopted (此处"实现"似应为"实行"——译注) catch-up strategy, and they all aimed at importing edge-cutting technologies that cannot be utilized? To this question, advocators to comparative advantage strategy theory may respond this way: it is because of the big factor endowment gap between developing countries and developed countries that the technologies suitable to the factor endowment of developed countries are not suitable to the factor endowment structure of developing countries. Then comes the second question: why cannot developed countries develop technologies suitable to the factor endowment structure of developing countries and sell them to developing countries (since they are able to develop the most edge-cutting technologies)?

Obviously, the question is not so simple as stated by the comparative advantage strategy theory. The logic of comparative advantage strategy theory "technological structure is endogenous in factor endowment" has surely neglected some important factors in technological progress. Because much experience has shown that technological structure cannot be improved with the improvement of the factor endowment structure. Even with the "former Soviet Union economy" on which the comparative advantage strategy theory is grounded, during 1960–1989, the investment rate in the Soviet Union was 29 % of its GDP, as compared with only 21 % in the United States. The analysis conclusion on the Soviet Union shows that, the result from the massive capital accumulation was that the productivity was almost not increased, and as the productivity was not increased, the economy would inevitably come to stagnation at the end. Furthermore, just as Lin (2002) also pointed out that, "their view (the 'suitable technology' theory) cannot explain why the efforts by the governments of countries in Latin America, Africa and Asia except the Four Little Dragons to increase saving rate were unable to increase the economic growth rate". All these examples indicate that improvement of factor endowment structure does not means sure upgrading of technological level.

Moreover, comparison of technological levels between countries with similar factor endowment structures can better explain this fact. Table 6.3 shows the calculation of productivity ratios of some countries with similar factor endowment structures. According to Table 6.3, Canada and the United States are almost equal in per labor physical capital and human capital, therefore, Canada and the United States are roughly at the same level in factor accumulation. However, by comparing their productivity, we found that Canada is only at 77 % of the level of the United States, therefore it is obviously poorer than the United States. Finland and the UK are roughly equal in per labor output level, however, the per labor output level of Finland mainly depends on factor accumulation, while the UK has higher productivity. Kenya and Tanzania have roughly identical factor accumulation, but the productivity of Kenya is over two times that of Tanzania, therefore its labor output is much higher than that of Tanzania.

We can also have a look at the microscopic experience of enterprises, Polanyi (1962) described two entirely different experiences in two countries in using machines to make light bulbs. In the 1950s, Hungary imported this type of machine, but they were unable to make a bulb without flaw in a whole year. In the same period, the same type of machines operated well in Germany. In the 1960s, the Cummins Engine of America respectively set up joint venture companies in Japan and India, to produce the identical truck engines. The factory in Japan quickly reached the quality and productivity level of the company. But in the factory in India, the cost was 2 times higher than that in the American factory, but the product quality was low. Polanyi (1962), Acemoglu and Ziliboui (2001) believed that in

Table 6.3 Productivity ratios of some countries with similar factor endowment structures

Country	Per labor output (y)	Per labor physical capital (k)	Per labor human capital (h)	Comprehensive production factors ($k^{1/3}h^{2/3}$)	Productivity (A)
United States	1.00	1.00	1.00	1.00	1.00
Canada	0.76	1.02	0.98	0.99	0.77
Finland	0.71	1.14	0.89	0.96	0.74
United Kingdom	0.70	0.80	0.82	0.81	0.87
Kenya	0.04	0.02	0.53	0.18	0.23
Tanzania	0.02	0.02	0.45	0.16	0.09

Source Per labor output: Heston et al. (2002); *physical capital* Bernanke and Gurkaynak (2001); *education* Barro and Lee (2000), the data here are continuous data for the period of 1960–1998

these cases, the critical point lied in the different practical experience of main engineering and management personnel, instead of the quantity and quality gap of physical capital.

Since the observed experience does not comply with the logic of comparative advantage theory "technological structure is endogenous in factor endowment", does it mean that the logic of comparative advantage strategy theory "technological structure is endogenous in factor endowment" has neglected some important factors in technological progress? In fact, in the transformation of economy to quality oriented economic growth stage, the growth theory made new explanation of the sources of sustained growth of economy, stating that it is technological progress instead of material capital that is the source of economic growth, and according to the endogenous growth theory, technological progress depends on human capital, instead of material capital. However, in the comparative advantage strategy theory, the decisive role of human capital in technological progress is weakened because the decisive role of material capital is emphasized. Of course, it is not deniable that necessary material capital should be allocated to allow the human capital to play its role, but the key to the question is: in the stage of world economy shifting to quality oriented economic growth, human capital has a more decisive role than material capital in technological progress and upgrading of industrial structure, and this point has been generally accepted in researches on the contribution rate from human capital. To this point, vindicators of comparative advantage strategy may respond this way: "creating advanced production factors (such as human capital) will surely require massive investment, as investment can only come from the economic surplus created by enterprises and the whole economy in the past production activities, there will be sufficient surplus to conduct human capital investment only by following the comparative advantage strategy." To this response, Schultz has in fact made the best reply, according to his calculation, during 1900–1957, the investment of the United States in "materialization" increased by 45 times, while the education investment increased by 85 times, however, the profit created by "materialized" investment increased by only 35 times, as compared with 175 times in the profit created by education investment. Since the contribution from human capital investment is so high, why cannot the gains from human capital investment be used to make up the cost of material capital invested for fostering human capital, and this must be done under the condition when this cost is very low? Of course, there are many factors that are related to the technical selection in enterprises, not merely the human capital level, for example, the intellectual property right protection system, size of markets, hindering by vested interest groups and the availability of tacit knowledge.

Next, in the relations between industrial structure and factor endowment, the comparative advantage strategy theory believes that industrial structure is endogenous in factor endowment also based on the same ground that technological structure is endogenous in factor endowment structure. In the stage that economy transforms to the quality oriented economic growth, technology and knowledge industrial structure also gradually becomes pillar industries, however, because the development of technology and knowledge industries depends on technology and knowledge, while technology and knowledge in turn depends on high quality human capital, entrepreneurship and associated institutional environment, etc. instead of capital factor. Therefore, when the conditions of high quality human capital, entrepreneurship and associated institutional environment (such as intellectual property rights, general market capacity and hindering by vested interest groups) cannot be satisfied, technology and knowledge cannot be readily obtained, therefore it is not possible to realize upgrading of industrial structure. Naturally, the comparative advantage strategy theory is unavoidably questioned a lot, such as "if transformation of leading industry cannot be realized (by developing economy according to comparative advantage strategy), is that conducive to the transformation of comparative advantage?", and "comparative advantage strategy has neglected dynamic trade interests such as readjustment of industrial structure, technological progress and institutional innovation, (therefore) developing countries developing international trade by following comparative advantage strategy have not only failed to narrow their economic gap with developed countries, on the contrary, they have fallen into the comparative advantage trap".[32]

3. Can economic convergence be realized by following the comparative advantage strategy?

Can the developing countries realize economic convergence with developed countries by following comparative advantage strategy? In fact, the answers to the previous two questions have provided ready answers to this question. It is as Lin (2002) said, "the per capita revenue of a country is a function of its technology and industries", "the root cause of developed countries being rich lies in their advantages in industries and technologies". So, developing countries should narrow their gap with developed countries in technology and industrial structure in the course of economic catch-up, and this must be realized by enterprises with viability. In the stage of economy transforming to production factor application efficiency driven economic growth, the improvement of factor endowment structure (K/L) will not necessarily lead to rise of viability of enterprises engaged in making technology and knowledge-intensive products, nor lead to the upgrading of technology and industrial structure, therefore, developing countries following the comparative advantage development strategy, under the condition that the world economy transforms to the quality oriented economic growth stage, cannot necessarily realize economic convergence with developed countries.

In the stage of economy transforming to quality oriented economic growth, the determinant of enterprise "viability", along with the in-depth development of knowledge and technologies, transformed from depending on the comparative advantage of factor endowment to the new stage of depending on scientific and technological progress, and the scientific and technological innovation ability has also become a determinant in the competitiveness of a country. Developed countries maintain their leading positions by means of their own technological advantages, they control markets and resources with technologies, to form high degree of monopoly on world market, especially on high-tech market, and they also obtain excessive monopoly profits with their technological advantages in technology and knowledge-intensive industries, products and production areas. The developing countries have no technological advantages, so they can only sell the low priced labor and sacrifice their own environment and resources for very little processing earnings with their comparative advantage in labor resources. This division of labor pattern determined on the basis of technology—labor resource comparative advantage makes the prospects for developing countries to realize economic catch-up not optimistic, mainly because the division of labor pattern determined on the basis of technology—labor resource comparative advantage has produced two effects: (1) this division of labor pattern has allowed the developed countries to produce more endogenous technological progress in the "learning by doing" research and development, and such advantage can be enhanced with the lapse of time, which will lead to eternal gap between the north and south in economy and technology, and the inability of the southern countries (developing countries) to realize economic catch-up; (2) the self-reliance research and development of technologies has the role to increase innovation ability and absorption ability, if developing countries mainly engaged in labor intensive industries fail to pay sufficient attention to strengthening their self-reliance research and development to raise their ability to imitate foreign technologies, they cannot increase innovation ability through self-accumulation of knowledge products, therefore technological catch-up will be out of the question. It is because the top ten countries with the United States, Japan and Germany taking the leader are in possession of 84 % of the research and development resources of the whole world that the technological and economic positions of all countries in the world have not changed much over more than half a century.

4. The logic source of unsuitability of the comparative advantage strategy theory

Following the analysis pattern of "factor endowment—viability—technology and economic structure selection—economic growth (convergence)" of comparative advantage strategy theory, both its internal logic system and its consistency with the realistic experience indicate the unsuitability of the comparative advantage strategy theory in the production factor application efficiency driven economic growth stage. To investigate its source, in the quality oriented economic growth stage, the determinant of enterprise viability is no longer the factor (capital) cost, technologies and industrial structure are no longer merely endogenous in factor endowment structure (K/L), therefore in the stage when economic catch-up can be realized only by depending on the improvement of application efficiency of production factors, it may not be possible to realize economic catch-up by following the economic development strategy of factor endowment comparative advantage. Although the comparative advantage theory has taken into account the role of high quality factors, but its logic is that, the increase of enterprise viability and the upgrading of technologies and industrial structure will need high quality factors, but fostering high quality factors needs capital investment, and so it further believes that the most exogenous variable in economy is the capital endowment. It has just neglected that capital is only a tool to produce high quality factors, instead of the high quality factor itself. And as a key to determine if the economy is convergent, high quality factors not only depend on capital, and the return from high quality factor can also make up the capital cost invested to foster it. If the negligence of the comparative advantage strategy theory made this theory suitable to the initial period of industrialization because of the critical role of capital in the quantity-oriented economic growth stage, in the transformation of economic growth pattern to the quality-oriented pattern, this should be the root source of all questions on this strategy theory, because only these high quality factors are critical to realizing the transformation of economic growth pattern and the economic rising, therefore we can say that the comparative advantage strategy theory is suitable to the early and medium periods of industrialization, but in the late period of industrialization or in the process of new industrialization, this theory has shown some unsuitability.

6.2.5 Multiple Characteristics of Large Country Economy and Unsuitability of Comparative Advantage Strategy Theory

Realizing economic growth with increasing earnings by relying upon human capital as well as technological progress, to increase the contribution from technology and knowledge intensive industries to economic growth, is the key for China to realize transformation of economic growth pattern at present. However, there are obvious multiple characteristics in the development of regional economy and industrial development in China, therefore, the economic development strategy of China faces multiple choices in the course of promoting the transformation of economic growth pattern.

First, the Large Country characteristics of China have made it possible to more effectively upgrade its industrial structure. According to the comparative advantage theory, it is necessary to develop technology and knowledge intensive industries under the condition of overall improvement of factor endowment structure, however, "any inducement of production factor intensity transformation triggered by technological progress will be relentlessly disintegrated by the fact that China has massive inexpensive labor forever and there is extremely abundant labor factor. … resulting in the rigidity of transformation of industrial structure".[33] Different from small countries, Large Countries, because of comparatively large overall scale, can concentrate part of resources to access into industries with higher factor endowment requirements, thereby greatly shortening the time of upgrading the whole industrial structure. Also, they have big domestic markets, which enable them to realize economies of scale by utilizing the "home market effect" so termed by Krugman, to develop key technology and knowledge intensive industries. Therefore, the "Large

Country" characteristics of China have provided conditions to "realize leap-forward development with self-reliance innovation in important areas" and develop key technology and knowledge intensive industries, which is not a question that can be solved merely by the comparative advantage strategy theory.

Next, according to the multiple characteristics of Large Country, different regions should have distinctive targets and routes for the upgrading of different industries and technological structures. By making use of differences in regional factor endowment, there can be a wider range for industrial selection, to push forward the upgrading of industrial structure by levels. According to the comparative advantage strategy theory, if we put the total resource endowment conditions of the whole country into labor intensive comparative advantage industries, the upgrading of industrial structure will surely be slowed down because of negligence of the economic diversity characteristics. If all regions aim at developing capital intensive and high and new technology industries, it is only grasping the shadow instead of the essence in the relative backward central and west regions and small and medium-sized cities. In the meantime, it goes without saying that the routes of technological and structural upgrading should be different for high-tech industries and traditional factor intensive industries, as Lee et al. (2005) pointed out that, in considering the technological progress in industries, analysis should be made on a case by case basis in conjunction with the technological characteristics of industries. Factors related to the technological progress of industries include market factors, technological characteristics of industries and technological tracks and capabilities of industries, and these factors have different effects and roles in the technological structure upgrading in different industries, not all depending on the most exogenous variable of the comparative advantage theory—the factor endowment structure.

Further, in view of the multiple characteristics in regional economic development, China is a Large Country with unbalanced regional economic and social development, with great differences in the construction of market economy system and in factor endowment structure, technological level, self-reliance innovation ability and demand conditions, different regions are in different development stages of the progressive economic transition, and the "diversification" of economic development level is a general picture of the different regional development in China. In the old industrial base of northeast, the state-owned economy lacking in viability makes up an important part (the institutional transition has not yet completed and the market economy system has not yet fully established), the development strategy in this region should be mainly encouraging enterprise transformation and rational allocation of factors; in the broad central and western regions, the economic development is in the extensive growth stage that economic growth is promoted by increasing factor input, with quite high potentials to promote economic growth by deepening capital and increasing the quality and quantity of factors input; in a small number of developed coastal areas, capital has been further deepened, the effect of diminishing marginal return of factors has become prominent, and the sustained economic growth is daily more depending on technological progress. This reality of "diversified" economic development level is a general picture of the staged differences in regional economic development in China. Therefore, the economic development strategy of China should embody the differences of regional economic development stages in the country, for example, different regions can adopt different development strategy layouts according to their respective development stages, to obtain complementation in differential development, and finally form the optimized region and country win-win development strategy pattern. However, the attention paid by the comparative advantage strategy theory to this issue is far from enough.

During reform and opening up, China implemented a development strategy oriented by comparative advantage, greatly stimulating the rapid growth of Chinese economy and raising the position of China in world economy. However, with the transformation of economic growth pattern, the comparative advantage strategy has received increasing questioning, then, is the comparative advantage strategy suitable to the transformation of economic growth pattern in China? The author is of the opinion that, as the economic growth pattern shows a staged transformation from "factor input quantity driven" to "production factor application efficiency driven", the comparative advantage development strategy is suitable to the "factor input quantity driven" economic growth pattern stage, because in this economic growth stage, capital plays a critical role. Then, "developing countries should adopt the development strategy of following the comparative advantage of their countries to realize the goal of transforming from an agricultural country to an industrial one".[34] But in the transformation from "factor input quantity driven" to "production factor application efficiency driven" stage, according to the analysis pattern of "factor endowment—viability—technology and economic structure selection—economic convergence" of comparative advantage strategy theory, it is unsuitable both in its internal logic system and on its consistency with the realistic experience. To realize the transformation of economic growth pattern, we can at least conduct more in-depth exploration on the following aspects for the issue of selecting economic development strategy.

(1) In the environment of contemporary world economic competition, it is usually difficult to realize the effect of comparative advantage strategy in developing countries. This is because: first, to give play to the comparative advantage, a country needs a factor price structure that can reflect the relative scarcity of production factors, i.e. a competitive market system, but market systems are not well developed in developing countries. Second, changes in international environment have changed the prerequisite to apply the traditional comparative advantage theory, the original assumed conditions and economic conditions are completely static, but the realistic production factors and resources can flow, breaking the form that developing countries participate in international division of labor with the comparative advantage of their resources, therefore it is surely difficult for them to narrow the gap with developed countries. Third, an important cause that it is difficult for developing countries to realize their

comparative advantage is the lacking of competitive advantage. Comparative advantage is a static advantage determined by the resource endowment and transaction conditions of a country, and a condition to obtain competitive advantage; competitive advantage is the action result of converting the potential advantages into comprehensive capability of realistic advantages, and the comparative advantage, as a potential advantage, can become a real export competitiveness only when it is finally converted into competitive advantage.

(2) In the transformation of economic growth pattern, the comparative advantage strategy theory is not suitable to the production factor application efficiency driven economic growth stage, in the quality oriented economic growth stage, the determinant of viability is no longer factor (capital) cost, because in the trend that the world economy transforms to the production factor application efficiency driven economic growth stage, the determinant of enterprise viability has also changed, from the factor cost to the application efficiency of factors. In the factor application efficiency driven economic growth stage, the TFP contribution in macroscopic gross quantity can exceed that from factors only when the contribution from factor application efficiency to the sales income of an enterprise exceeds that from factor input quantity.

(3) Currently in the economic development of China, it is necessary to well combine the labor resource advantages with the technology and knowledge intensive industries. The comparative advantage of labor force resource endowment in China still has great effect on increasing employment and promoting economic growth, so it must be utilized in the present stage. However, to realize the transformation of economic growth pattern, great efforts must be made to increase the role of scientific and technological progress, and to realize "breakthroughs at main points" in technology and knowledge intensive industries and critical sectors. The comparative advantage development strategy can really well explain the success of Chinese economy in reform and opening up, but one important factor was neglected, that is, China is a Large Country with very great differences internally. In either Japan, or the "Four Asian Tigers", their internal differences in economy and resource endowment cannot be compared with those of China. A "comparative advantage" embodied by factor endowment obviously cannot fully summarize the "Large Country" characteristics and the "diversified" economic and technological characteristics of China. The comparative advantage of resource endowment is not the necessary and sufficient condition of the industrial competitiveness and enterprise viability of a country, but lacking in technological advantage has made it difficult to continue with the comparative advantage of the labor intensive industries in China. In international economic division of labor, China has remained over a long time in the labor intensive industries and at the low end of value chain, lacking in competitive advantage and controlled by others in critical technologies, the foreign-funded enterprises have daily increasing control over the main industries and markets of China. Although China is the third biggest country in world trade, the contribution

from foreign-funded enterprises is nearly 2/3, in 2006, the import and export of foreign-funded enterprises accounted for 58.9 % of the total foreign trade volume of China, and the foreign trade dependence was as high as 63.8 %. The economic sovereignty of China has gradually lost with the deepening of opening to the outside world, and this stern economic situation calls for a new Large Country development strategy to raise the core competitiveness of the country. Therefore, China, as a rapidly rising big economy country, cannot simply follow the industrialization experience represented by "the Four Asian Tigers", instead, it should take a road of development of a Large Country, therefore, we should, according to the "Large Country" characteristics and "diversified" characteristics in economic development of China, make further systematic exploration on how to choose suitable economic development strategies (for example, how different regions should position their functions and adopt different development strategy layout based on their respective development stages, how to complete the inter-regional coordinated and linked development mechanism to realize coordinated development of regional economy, the target and route of upgrading different industrial and technological structures and the corresponding policy supports), and this is an issue with major strategic realistic significance and important theoretical significance in the transformation of growth pattern in China.

6.3 Economic Development Strategy Oriented by Comprehensive Advantages of Large Country

A basic judgment of the national conditions is the prerequisite for selecting the economic development strategy. China has changed from a Large Country of agricultural economy to a big industrial country, which means that the modernization of Chinese economy has come to the new stage of realizing the transformation from a Large Country of industrial economy to a powerful country of industrial economy, with pushing forward the industrial modernization process as the core mission. The main task in this stage is to realize the transformation of economic growth to the intensive growth pattern, and the transformation and upgrading of industrial structure from resources-intensive and labor-intensive to capital-intensive and technology-intensive industries, and establishing the advanced industry technology supporting system, and this is the key to realizing the transformation of economic growth pattern and the rising of Large Country economy. The industrial structure strategy is critical to economic development strategy, the foreign trade strategy, science and technology development strategy as well as regional economy development strategy should all adapt to and follow the basic orientation of industrial structure strategy. The orientation of industrial structure strategy depends on the existing and potential advantages of an economy, therefore, the economic development strategy based on comprehensive advantages of Large Country should embody the characteristics of a Large Country and its multiple economic and technological characteristics.

6.3.1 Development Strategy Oriented by Large Country Characteristics

1. Foreign trade development strategy should embody the Large Country characteristics

A very big difference between Large Countries and small countries is that Large Countries can govern the prices of products on international market, while small countries can only accept prices on international market. The prices as low-price competition among the labor intensive industries in late developing Large Countries are very close to cost, with only limited space of price adjustment. However, to developing small countries, even if the trade conditions deteriorate, they can mitigate the detrimental effect produced by deterioration of trade conditions of some categories of commodities by regulating the export structure and diversified production, but it is much more difficult for Large Countries to regulate export structure than small countries, because Large Countries can influence and even determine the international price in quite many industries, and large total resource quantities are required for Large Countries to upgrade their industrial structure. According to typical neoclassical growth theory, the rise of capital to labor ratio reduces the marginal contribution of capital inventory to output, therefore growth is subjected to restriction by diminishing marginal return of factors. Large Countries are more inclined to the "anathema of diminishing marginal return of factors", because as long as a Large Country increases the supply of one product, the price of that product on world market will be decreased. Therefore, the increase of output in Large Countries will reduce the value of output through the deterioration of trade conditions, amplifying the effect of factor accumulated diminishing marginal return. Also, it is far more difficult for a developing Large Country than a developing small country to get rid of the "comparative advantage trap" and "immiserizing growth" characterized by long-term low-added value and severe trade conditions. Also, according to the view of "fallacy of composition theory": one country can be successful in adopting a strategy, but when many countries adopt the identical strategy, all countries will fail. When developing countries such as India, Brazil and Mexico with similar comparative advantage also implement free trade and actively participate in international division of labor and export industrial finished products in large quantities, it will be more difficult to improve the trade conditions for the labor-intensive industrial finished products, the competition among labor-intensive products will become daily severe, and the probability of fallacy of composition will increase substantially.

In the actual situation of foreign trade development in China, the stimulating policy focused on giving play to the comparative advantage of labor and resource endowment and "earning foreign exchange through export" was really successful in nearly 30 years of reform and opening up in China. However, in the meantime, guided by this foreign trade strategy, China has formed a foreign investment pattern dominated by processing trade, as demonstrated by: in 2006, the total processing trade import and export volume of China was 831.9 billion US dollars, accounting for 53 %; enterprises funded by foreign capital and Hong Kong, Macau and Taiwan capital became the main body to push the rapid growth of foreign trade export from China, in 2006, in the foreign trade export from China, export from enterprises funded by foreign capital and Hong Kong, Macau and Taiwan capital was as high as 58.18 %, in 2007, the proportion of enterprises with sole foreign investment in the total processing trade import and export volume of the whole country approached 64.8 %, although high-tech products with computers, micro-electronic and mechanical and electrical products as the main have accounted for a high proportion, because of lacking in core technologies, China remains in essence in a typical development stage of simple processing and assembling. Most processing trade not only is low in technical content, no domestic supply has been realized as substitutes for many intermediate inputs, therefore the value-added ratio is low, with an apparent feature of "extracorporeal circulation", the processing trade could only give very little drive to other associated industries and those in the medium and upstream sectors in the country, making it difficult to upgrade the industrial structure with the technological spillover effect, as a result, the foreign investment pattern with processing trade as the main has been frequently questioned. Also, as most of the commodities exported from China are labor-intensive products and ordinary mechanical and electrical products with little demand elasticity, the rapid expansion of export trade of China will surely lead to reduction of export price, while China mainly imports capital, energy and technology intensive products, which are at rigid prices, and this inherited structure of import and export trade results in deterioration of price and trade conditions. Furthermore, other developing countries also mainly export labor intensive products, leading to intense competition. Therefore, the trade conditions of China have been in the trend of deterioration since the 1980s. In 2007, the total import and export trade volume of China exceeded 2 trillion USD, steadily ranking at the 3rd place in the world, with annual trade surplus of 262.2 billion US dollars. Large amount of foreign investment also flowed in, so China already has quite abundant foreign exchange reserve. The stimulating foreign trade strategy with "earning foreign exchange through export" as formulated in the situation of foreign exchange and capital shortage is no longer suitable to the transformation of foreign trade growth pattern in China today with the increase of capital abundance. To realize the transformation from a big trading country to a powerful one, we should take the road of Large Country development.

First, we should give play to the scale advantage of domestic market of a Large Country, and combine domestic and foreign trade to reduce the external economic risks. China depends on foreign trade too much, and the dependence was as high as 70 % in 2007, implying high economic risks. Facing the increased foreign trade uncertainty, to properly resolve the potential risks resulted from high foreign trade dependence and increase the ability of domestic economy to withstand and digest external economic risks, we must expand domestic demand to maintain the stability and sustainable development of economy. For economic growth, we should implement a Large Country economy strategy of balanced development combining internal and foreign trade and coordinating overseas demand with domestic demand. While developing foreign trade, we should take expanding domestic demand as a long-term strategic policy and supporting point in economic development.

Next, we should, at a strategic level, establish the import mechanism of advanced equipment and technologies, especially strategic resources, push forward the joint procurement mechanism of bulk resource products, embody the import scale advantage of a Large Country to the maximum extent, and enhance the "right of speech" of China in foreign negotiations.

Furthermore, we should substitute the extensive traditional export pattern with the quality growth pattern characterized by technological progress and increasing efficiency. At present, the foreign trade of China is in a critical period of transforming from quantity-oriented growth to quality and efficiency based growth, and from a big trade country to a powerful trade country. The central link in transforming the foreign trade growth pattern is technological progress, therefore we should make efforts to create an environment suitable to the technological innovation and self-reliance research and development by Chinese enterprises, to raise the self-development capacity of these enterprises. In the division of labor and trade within industries, we should upgrade our products; and in the division of labor and trade within products, we should upgrade from low end to high end of the value chain.

Finally, we should adjust as appropriate the policies for foreign investment, to promote the transformation from "quantity-oriented" to "quality and efficiency oriented" in the utilization of foreign investment, and on the basis of unifying the tax system for both domestic and foreign invested enterprises, preferential policies can be implemented in key areas to attract foreign investment on a selective basis, in the meantime, supervision over foreign investment should be strengthened and foreign investment performance assessment system be established, to improve the quality of utilizing foreign capital. We should gradually change the development pattern of attracting foreign investment with preferential policies and cheap resources advantages, driving export with foreign investment and driving economic growth with export.

2. Large Country characteristics can provide more choices for industrial development strategies

According to the development strategy of comparative advantage, industrial structure upgrading depends on the dynamic changes of factor endowment structure. The comparative advantage strategy theory states that, if under-developed countries choose to follow the strategy of comparative advantage, then the factor endowment structure can be upgraded very quickly. The "Four Asian Tigers", and even Japan in bigger scale, smoothly realized upgrading of industrial structure by improving their factor endowment structure. However, China, as a developing Large Country, has its special conditions, it is lacking in capital but has abundant labor, and more important, due to the existence of the dual economic structure, with the industrial restructuring and completion of market economic system, massive labor force moved out of rural areas, so that the labor intensive comparative advantage industries in China have almost unlimited supply of labor force, which makes it very difficult to improve the factor endowment in China in a short period of time. Just as pointed by Yu Weihua and Qin Botao, wages is the most important index to measure the factor endowment structure of capital to labor ratio, after the 1980s, there was no obvious increase in the actual wages of migrant rural workers in China, and this means that, after rapid development for nearly three decades in China, after massive accumulation of capital, the resource endowment in China has not been fundamentally changed, and this has resulted in difficulties in industrial structure upgrading in China. Therefore, to a country with fairly small scale of economy, adopting the comparative advantage development strategy can undoubtedly achieve the highest speed of capital accumulation, thereby upgrading the factor endowment structure and finally upgrading the industrial structure. However, Large Countries cannot effectively upgrade their industrial structure with this development strategy. In fact, the Large Country characteristics of China can provide more choices for industrial development strategies.

First, a Large Country can have a wider range of industrial selection. According to the analysis framework of comparative advantage theory, with countries as decision-making units, if two countries, one small and one big, have the identical factor endowment degree, then these two countries should follow the same route to regulate their industrial structures. But in fact, the production possibility curves of these two countries are not identical, because many industries have the access threshold of minimum factor endowment. The industrial structure of small countries will gradually transform from labor-intensive to capital-intensive and technology-intensive with the increase of the overall factor endowment degree. However, Large Countries with comparatively large overall scale can concentrate part of resources to access into industries with higher factor endowment requirements, thereby greatly shortening the time of upgrading the whole industrial structure. Therefore, to realize the transformation of economic growth pattern, China can concentrate resources in technology and knowledge intensive industries and critical sectors, to realize development in priority and breakthroughs at main points.

Next, Large Countries can adopt strategic industrial policies to promote industrial structure upgrading. Government can, based on its CAOLC, support the technology-intensive industries in a short period of time, to promote them to improve technology, and obtain the benefit of economies of scale and learning by doing. Furthermore, the domestic market advantages of Large Countries can lay some foundation for their strategic industrial policies, thereby guaranteeing the effective optimization and upgrading of the industrial structure in their countries. Late-developing Large Countries need not go into international division of labor according to comparative advantage in a totally passive manner, instead, they should well combine the utilization of comparative advantage with fostering new competitive advantage, to actively obtain advantages through technological progress and human capital investment, so as to change the general comparative advantages in a country and promote the upgrading of industrial structure.

3. Large Country characteristics make it possible to select scientific and technological development strategy aiming at breakthroughs at key points

The key to economic rise and transformation of economic growth pattern in Large Countries is establishing an advanced industrial technology supporting system and providing a leading industrial structure. The market capacity of Large Countries can attract FDI from transnational companies, thereby helping giving play to later-mover advantage in technologies by channels such as technological overflow, however, it is not possible to establish an advanced industrial technology supporting system only by giving play to the later-mover advantage in technologies. Possessing the self-reliance technological innovation capability and proprietary intellectual property rights in strategic and basic technological fields is the key to realizing transformation of economic growth pattern and economic rise in China. Moreover, the market capacity of Large Countries can increase the expected gains from innovation in enterprises, therefore conducive to realizing the innovation advantage. The gross resource advantage of Large Countries also makes it possible to implement the scientific and technological development strategy aiming at breakthroughs at key points. This also requires government to concentrate resources to organize and support on a selective basis some basic researches and major subjects for technological breakthroughs; to establish self-reliance innovation system and mechanism in critical sectors; to establish industrial technology progress system, and form a government—enterprise—research cooperative research system; to determine the strategic key technologies and their industries, and adopt specific supporting countermeasures; to guide and support key technologies with policies and regulations; to select state-owned large and medium-sized enterprise with innovation potential and support them in both capital and technology, to give play to their capital (including human capital) and technological advantages with innovation; and to select key technological fields conducive to implementing catch-up strategy as the main part of catch-up plan and scenario, and support mainly the development of leading industrial technologies and key technologies that can realize technological catch-up, until they reach the international leading position in technology.

6.3.2 Development Strategy Oriented by "Diversified" Characteristics

The diversification of regional structure, economic structure and technological structure can evolve to diversification of development strategy, motivation, main players and industries, the economic multiple characteristics of Large Countries are the basis to form the CAOLC, and also provide coupling space for Large Countries to adopt composite development strategies.

(1) On industrial and foreign trade development strategy, according to the economic multiple characteristics, the industrial selection range can be expanded by using the differences in regional factor endowment, to push forward the upgrading of industrial structure and trade structure by levels. After reform and opening up for nearly 30 years, the developed east coastal region of China has gathered massive material capital and human capital, the factor endowment has changed greatly, and technological level also been raised substantially, having acquired to a certain extent the conditions to develop capital-intensive and technology-intensive industries, therefore, the developed eastern region can be first encouraged to upgrade its industrial structure, furthermore, between the east coastal regions and central and west regions, and between extra-large cities and small and medium-sized cities in China, there are not only great differences in their factors such as labor, capital (including human capital) and technology, their development extent and completion of market institutions and systems also differ greatly. Therefore, by following the theoretical standard of comparative advantage, there should be different targets and routes in these regions to upgrade their industrial structure. If all regions of the whole country all upgrade their trade structure with no regard to practical conditions, it will probably achieve little with great efforts. Also, if we put the total resource endowment conditions of the whole country into labor intensive comparative advantage industries, the upgrading of industrial structure will surely be slowed down because of negligence of the economic diversity characteristics. Therefore, the upgrading of industrial structure in China must be founded on the reality of "diversified economy", and be pushed ahead level by level. For example, we can realize the space restructuring of advantages of the east region and central and west regions through division of labor between regions. In the division of labor system between industries, the east region can, at the appropriate time, transfer its labor intensive industries to the underdeveloped central and west regions, to develop in priority capital-intensive and technology-intensive industries; in the division of labor system within industries, the east region can give play to its advantages in capital and technologies, to focus on developing high quality products, while the central and west regions can bring into play their advantages on labor and land resources, to develop ordinary products; in the division of labor system within products, the east region can be mainly specialized in the high end links of value chain such as research and development, marketing and management, while the central and west regions can be mainly specialized in the labor-intensive processing sectors. With this space restructuring of comparative advantage and cooperation in industrial division of labor in different regions, it can be expected that the developed east coastal region can, through self-dependent innovation, develop capital and technology intensive industries, and extend and even strengthen the traditional comparative advantage of China in the

under-developed central and west regions through technological transfer and industrial transfer, and finally realize the "hierarchic upgrading" of the overall industrial structure in the whole country.

(2) The technological diversified characteristics of late-developing Large Countries call for diversified technological progress pattern in China. Industries and regions with high technological level can adopt the technological progress pattern led by self-reliance innovation; industries and regions where the technological level is not quite high should be encouraged to make creative imitation of imported technologies through "reverse engineering"; and industries and regions with big gap in technology should be encouraged to raise their imitating ability of foreign technologies by way of self-reliance research and development and strengthening human capital accumulation. In this way, different sectors in various regions can adopt different "hierarchic and diversified" leading technology progress patterns, strengthen self-reliance research and development while giving full play to the technology later-mover advantage, to continually increase the ability to absorb, imitate and innovate on foreign technologies, so as to realize technological progress in a Large Country. This technological progress can on one hand promote upgrading of factor endowment and technological accumulation, to allow a late-developing Large Country to rank in higher links of international division of labor; and on the other hand, it can foster advanced and specific factors, so that the comparative advantage embodied by factor cost can last long and develop in favorable cycles.

(3) In the development of regional economy, on the basis of the diversified characteristics, it is necessary to form a favorable pattern with rational division of labor and complementation of advantages as oriented by target diversification. This requires us to correctly understand the economic advantages in different regions and correctly handle the relations between potential and realistic advantages and between resources advantage and other advantages, so as to select the economic development strategy that can give maximum play to the economic advantages in the region. In the meantime, in the industrial development in different regions, appropriate combination of regional specialization and necessary diversification should be adopted. It is necessary for the state to measure and calculate the comparative advantages and competitive advantages in various provinces and regions, to determine their advantageous industries, the leading industries of the state and layout of major development areas, and all regions should base themselves on their objective resource endowment differences, to form a rational inter-regional division of labor pattern and regional industrial structure with complementary advantages and disadvantages, to bring into full play the advantages of all regions. This will enable realizing an economy with division of labor: the comparative advantage can be brought into play through division of labor between industries, in addition, such division of labor can also create conditions for mechanized production and comprehensive utilization of resources, and economies of scale advantages can also be realized through division of labor within industries and within products.

In short, to realize transformation of economic growth pattern and economic rise in Large Countries, Large Countries with unbalanced economic and technological development should make necessary correction and completion to the comparative advantage development strategy theory according to the characteristics of Large Country and features of diversified economy, and the industrial development strategy, foreign trade development strategy, science and technology development strategy as well as regional development strategy should all embody the characteristics of Large Countries and multiple economy characteristics, to bring into play the comprehensive advantages of Large Country. They should strengthen the R&D in labor intensive industries while giving full play to the comparative advantages of labor intensive industries, and increase the added values of products; they should attach importance to cultivating the ability of self-reliance innovation and solve issues in critical technologies on a selective basis, to establish an advanced industrial technology supporting system; they should make use of the characteristics of Large Country to expand the scope of industries selection, make use of the strategic industrial policies to promote the development of critical pillar industries; make use of the diversity of economy and technologies and differences in regional resource endowment to push forward upgrading of industrial structure by regions, by industries and by levels, and generate new advantages through trans-regional resources cooperation and technological cooperation. This will enable realizing the organic combination of natural evolution of industries and technologies under the comparative advantage strategy with the self-reliance advancement under the catch-up strategy, to effectively promote the industrial structure upgrading and technological progress in Large Countries, and realize the transformation of economic growth pattern and economic rise in Large Countries.

Notes

1. Both articles by Prof. Li Daokui: *China's Development Needs Large Country Strategy* and *Six Basic Factors in Large Country Development Strategy*, analyzed the issues of Large Country strategy and Large Country pattern, refer to: Li Daokui: *Large Country Development Strategy*, Peking University Press 2007 Edition.
2. Jing Xueqing: *Economic Development Patterns of Large Countries and Main Supporting Points to Economic Growth in China*, *Shanghai Economic Review* 2000 Vol. 5.
3. Yu Weihua and Qin Botao: *Large Countries and Comparative Advantage Development Strategy*, *Prediction* 2006 Vol. 5.
4. The research team of Fudan University, with Lu Ming, Chen Zhao, Wang Yongqin and Zhang Yan as main members, has explored for long period on the development road of China, including China's road of reform and opening up, China's Large Country development road and their world significance. Mainly refer to: Lu Ming et al.: *Large Country Economy Development Road of China* (*Economy Volume*), Encyclopedia of China Publishing House 2008 Edition.

5. Prof. Yao Yang does not fully agree with the comparative advantage strategy theory proposed by Prof. Justin Yifu Lin, advocating to complete the comparative advantage strategy theory with the limited catch-up theory, refer to: Yang Rudai, Yao Yang: *Limited Catch-up and Development of Large Country Economy*, Essays of 2006 *Sixth Annual Conference of China Economics*.
6. [US] Robert J. Barrow: *Determinants of Economic Growth*, China Social Sciences Publishing House 2000 Edition.
7. Prof. Ren Ruoen has studies over a long period on the issue of total factor productivity, with particular attention paid to the empirical verification of total factor productivity in China, refer to: Sun Linlin and Ren Ruoen: *Estimation of Capital Input and Total Factor Productivity in China, World Economy* 2005 Vol. 12.
8. Justin Yifu Lin, Pan Shiyuan and Liu Mingxing: *Selection of Technologies, Institutional System and Economic Development, Economics* (*Quarterly*) 2006 Vol. 3.
9. Justin Yifu Lin: *Development Strategy, Viability and Economic Convergence, Economics* (*Quarterly*) 2002 Vol. 2.
10. Same as 9.
11. Same as 9.
12. Same as 9.
13. Domestic scholars have already noticed the limitation of comparative advantage strategy, and have started studies on its suitability or adaptability, refer to: Xu Yuankang: *Research on Unsuitability of Comparative Advantage Strategy in Economic Development of China, Reform* 2003 Vol. 5; Yu Weihua and Qin Botao: *Comparative Advantage Development Strategy in Large Countries, Prediction* 2006 Vol. 5; Guo Kesha: *Comments on Foreign Trade Strategy and Trade Policies of China, International Economic Review* 2003 Vol. 1.
14. Wang Yungui: *WTO and China's Trade Development Strategy*, Economic Management Press 2002 Edition.
15. The Economic Development Research Center of Wuhan University studied over long time on the later-mover advantage theory, and advocated to explain the rapid and sustained growth of Chinese economy with the later-mover advantage of developing countries.
16. Many empirical and theoretical researches support the regularity of economic growth stages, such as Forter (1990), Kuznets (1971, 1973), Moses Abramovitz (1993); Samuelson and Nordhaus (1992).
17. Zhang Ping and Liu Xiahui: *Front of Economic Growth in China*, Social Sciences Academic Press 2007 Edition.
18. Song Zexing and Fan Kang: *World Economic History* (*Vol. I*), Economic Science Press 1994 Edition.
19. Justin Yifu Lin: *Development Strategy, Viability and Economic Convergence, Economics* (*Quarterly*) 2002 Vol. 2.
20. Same as 19.
21. Because Justin Yifu Lin and his cooperation partners have made detailed expositions on the logics and basis of the proposition of comparative advantage strategy theory in many essays and works, we only make simple analysis here.

22. The role of technological progress has also been taken into account in the comparative advantage strategy theory, however, it emphasizes the importance of accumulation of material capital from the material conditions of technological progress, for example, "technological progress is possibly a relatively simple import, and is also possibly a self-reliance research and development research and development activity demanding large amount of capital, and both require input of capital. Therefore, technological progress is often restricted by the extent of capital accumulation" (Lin and Yongjun 2003). However, this part emphasizes that in the quality oriented economic growth stage, the role of technological progress is decisive in technology and knowledge intensive industries.
23. Science and Engineering Indicator, 1998, p. 52.
24. The role of human capital has also been taken into account in the comparative advantage strategy theory, as this theory emphasizes the role of material capital, it does not take human capital as a decisive factor in technological progress, instead, it emphasizes the importance of material capital accumulation from the complementarity of human capital and material capital. For example "the roles of human capital and material capital are complementary, increasing human capital solely without the matching of certain material capital with it, the high human capital cannot play its role, as a result, labor forces with high human capital will flow in large quantity to countries with high material capital, and this is the cause of 'brain drain' happening in many developing countries." (Justin Yifu Lin and Li Yongjun, 2003). However, this part emphasizes that in the quality oriented economic growth stage, the role of human capital is decisive to technological progress in technology and knowledge intensive industries.
25. Xu Jianbin and Yin Xiangshuo: *Deterioration of Trade Conditions and Effectiveness of Comparative Advantage Strategy, World Economy* 2002 Vol. 1.
26. Guo Kesha: *Comments on Foreign Trade Strategy and Trade Policies of China, International Economic Review* 2003 Vol. 1.
27. Hong Yinxing: *From Comparative Advantage to Competitive Advantage—Concurrent Discussion on Defects of the Theory of Comparative Advantage of International Trade, Economic Research* 1997 Vol. 6.
28. Xu Yuankang: *Research on Unsuitability of Comparative Advantage Strategy in Economic Development of China, Reform* 2003 Vol. 5.
29. Justin Yifu Lin: *Development Strategy, Viability and Economic Convergence, Economics (Quarterly)* 2002 Vol. 2.
30. About the understanding of technological structure, the comparative advantage strategy theory particularly emphasizes that a given technological level is linked with a given industrial section, for example "technological structure (or industrial section)" (Lin and Xifang 2003; Lin and Mingxing 2004). It is also said that "when the factor endowment structure of this economy is upgraded, enterprises are able to accordingly upgrade their products or technologies" (Lin 2002), "obviously, the existing technologies are formed by technologies of different levels from high to low" (Lin et al. 2006), and so on, from these expressions, we can easily see that the "upgrading of technological structure" of the comparative advantage strategy theory

is in essence the rise of technological level, or that the upgrading of technological structure will surely lead to rise of technological level.
31. Same as 29.
32. At the beginning of the 21st century, scholars in China started researches on the negative effect of comparative advantage strategy theory on developing countries, and made detailed analysis of the "comparative advantage trap" in foreign trade, which renders developing countries unable to catch up with and surpass developed countries forever, and remain in a passive position over long time in foreign trade. Refer to: Wang Diankai: *Comparative Advantage Trap and China's Choice of Trade Strategy*, *Economic Review* 2002 Vol. 2.
33. Yu Weihua and Qin Botao: *Large Countries and Comparative Advantage Development Strategy*, *Prediction* 2006 Vol. 5.
34. Justin Yifu Lin and Liu Mingxing: *Economic Development Strategy and China's Industrialization*, *Economic Research* 2003 Vol. 7.

Chapter 7
Case: Human Capital Advantages of Late-Developing Large Countries

In the previous chapters, we analyzed the characteristics, mechanism and strategy orientation of economic development in Large Countries. In this chapter, we will start case analyses. This chapter, with human capital as a case, analyzes in detail the human capital comprehensive advantages in late-developing Large Countries, and builds up an analysis framework based on system theory and coupling theory, to analyze the acting mechanism of human capital comprehensive advantages in late-developing Large Countries. In the meantime, it puts forth the assumption of high degree coupling of the heterogeneous human capital with diversified industrial structure, material capital investment and technological level, and made empirical verification with data of China, to propose corresponding policy recommendations.

7.1 Related Literatures

The relationship between human capital and economic growth has remained a focus of attention by economists since the 1960s. Theoretically, early theorists on human capital such as Schultz and Becker all emphasized that human capital including the knowledge and skills of people promotes production, and are main factors in economic growth. Arrow's learning model endogenized the "Solow residual value" into knowledge growth in "learning by doing", stating that knowledge and experience accumulation, as a production input, reduces the production cost of unit product with the increase of output, and in the meantime, the "spillover effect" of knowledge also increases the productivity of capital and labor. The catch-up mode of Nelson and Phelps (1966) broke the viewpoint to take human capital as direct input factor to the production process, stating that the human capital, through technological innovation and absorption, accelerates the spreading of technologies to promote economic growth. Romer introduced the R&D department into the growth model, to endogenize technological progress in the viewpoint of human capital, stating that technological progress is the result of endogenous accumulation

Y. Ouyang, *The Development of BRIC and the Large Country Advantage*,
DOI 10.1007/978-981-10-0633-3_7

of knowledge, while human capital is a key factor in knowledge accumulation, and human capital produces a positive "spillover effect" on economy by creating knowledge. Lucas (1988) even directly substituted "knowledge" with human capital, endogenized it, and by analyzing the "internal effect" and "external effect" of human capital, attributed the sources of economic growth in a country and the differences in the economic growth in different countries all to the human capital.

However, many scholars came to different conclusions in their empirical researches on the relations between human capital and economic growth. Mankiw, Romer and Weil (1992) used the proportion of people having received secondary education in the working-age population as a substituting indicator for human capital, and proved that the output elasticity of human capital is about 1/3. The researches by Barro and Sala-i-Martin (1995) show that, the expenditure on public education has significant positive effect on growth. Li Kunwang and Chen Lei (2005) conducted empirical analysis of the economic growth convergence of APEC during 1950–2000, and found that the average years of education is significantly positive-correlated to growth rate. The significant positive correlation between human capital and economic growth has also been proved by other scholars.[1] However, the transnational research by Pritchett (2001) shows that the increase of education level of labor force has almost no effect (and even has negative effect) on growth rate. Bils and Klenow (2000) also believed that human capital has no promoting role to output.

It is this inconsistency between theoretical analysis and empirical conclusions that made economists start paying attention to the research on the heterogeneity of human capital, i.e. to explore on the relations of human capital of different types and characteristics with technological progress or economic growth. Lucas (1988) classified human capital into human capital with general knowledge common in the society and the "specialized human capital" with special knowledge embodied by the work skills of laborers, and believed that only the latter is the "engine of economic growth." The evidences of Murphy, Shleifer and Vishny (1991) showed that, economic growth is relatively fast in countries with high proportion of college students majoring in engineering; while the economic growth is relatively slow in countries with high proportion of college students majoring in law. The theoretical model proof made by Grossman and Helpman (1992) on the role of skillful and unskillful labor in production has indicated that, the higher proportion of skillful labor in a country, the more rapidly its knowledge inventory and economy grow. Barro and Lee (1996) used sample data from 87 countries during 1965–1975 and from 97 countries during 1975–1985 to make regression analysis of the true per capita GDP growth rate, and found that the primary education level is not significantly correlated with growth rate, the secondary and tertiary education level of female has negative effect on growth rate, while the secondary and tertiary education level of male is significantly correlated with the growth rate. The empirical analysis made by Fedderke (2001) in South Africa showed that: different types of education and training contribute differently to economic growth. Among them, the science and technology education in tertiary education produces the most significant positive effect on total factor productivity. Young, Levy Daniel and Higgins

(2004) studied the effect of different types of human capital on economic growth by using the neoclassical growth model and data of the United States, and the result showed that human capital with the education level of bachelor degree and above is significantly positive-correlated to economic growth. Vandenbussche et al. (2006) found that it is the human capital having received tertiary education instead of average human capital that has significant promoting role to the total factor productivity. Domestic scholars also conducted in-depth researches on the heterogeneity of human capital. Ding Donghong and Liu Zhibiao (1999) classified human capital into homogeneous and heterogeneous forms. Li Zhongming (1999) classified human capital into four types on the basis of the competence types: the ordinary, the skill-based, the management-oriented and the entrepreneur-oriented. The research by Wang Jinying (2001) showed that, specialized human capital is a mark of productive force level, the main carrier of technological innovation and spreading, management innovation and implementation and institutional innovation, and the engine of economic growth. Chen Xiushan and Zhang Ruo (2006) made in-depth research on the contribution of heterogeneous human capital to the regional economy gaps, and the result showed significant differences in the contribution rate from junior, medium rank and senior human capital. Peng Guoping (2007) studied the relations between total factor productivity and human capital composition, and came to the identical conclusion as Vandenbussche et al. (2006), that is, only the part of human capital having received tertiary education produces a significantly positive effect on TFP.

In summary of the above, the research of the heterogeneity of human capital has become an important route to enrich and develop the human capital theory, and its basic conclusion is: human capital has an apparent feature of heterogeneity; human capital with higher education level or technical proficiency plays a more significant promoting role in economic growth or technological progress. This has not only proved the important promoting role of human capital in economic growth, but also provided basis for governments of all countries to select or determine the main area of human capital investment. However, the researches already made have the following restrictions: first, they laid particular emphasis on a single aspect, such as different education level and difference in competence, to study the effect of heterogeneous human capital on economic growth, therefore it is difficult to fully expound the role of human capital. In fact, the heterogeneity features of human capital include both quantity and quality aspects, therefore, it is necessary to take them into full account, to constitute a heterogeneous human capital system, and only in this way can it fully and effectively reflect the role of human capital in economic growth; second, methods such as regression analysis were most used to study if human capital play a promoting role in economic growth, and there were very few researches, especially empirical researches that were based on the realistic characteristics of late-developing Large Countries—the regional diversification, economic diversification and technological diversification, to reveal the important role of human capital in economic growth from the viewpoint of the adaptability of heterogeneous human capital to the diversified material capital investment, industrial structure and technological level, or the extent of matching between them; and third, because of the restriction by the "theory of human capital disadvantage in developing countries", few researches paid attention to the comprehensive advantage of human capital in late-developing Large Countries. The human capital in late-developing Large Countries is low, but their economy grows quite rapidly, and this phenomenon indicates that there is possibly a certain comprehensive advantage with the human capital in late-developing Large Countries, and this comprehensive advantage also enables the human capital at a fairly level to play an important pushing role in economic growth. As the diversified material capital investment, industrial structure and technological level in late-developing Large Countries require heterogeneous human capital to adapt to it,[2] therefore, analysis of heterogeneity and adaptability provide us with a new view angle to study the human capital comprehensive advantages in late-developing Large Countries. If the human capital at fairly low level in late-developing Large Countries can well match with the diversified material capital investment, industrial structure and technological level, it can also play an important pushing role in economic growth, i.e. demonstrating certain comprehensive advantages. We take human capital as a system containing both quantity and quality factors, to explore on the comprehensive advantages of human capital in late-developing Large Countries and its functioning mechanism in the view angle of the heterogeneity of human capital and its adaptability to the diversified industrial structure, material capital investment and technological level, and using the theory of system and the coupling theory in physics, and on this basis, we analyze the important role of human capital in late-developing Large Countries in economic growth, and put forth the policy recommendations on upgrading human capital investment suitable to the realistic national conditions of developing countries.

7.2 Analysis Framework and Evolution Mechanism

1. Analysis framework based on system theory and coupling theory

The Large Country characteristics and transformation characteristics of late-developing Large Countries have determined that they have comprehensive advantages integrating those of both developing and developed countries—comprehensive advantages of Large Country, the regional diversification, economic diversification and technological diversification are their outstanding realistic characteristics.[3] This has determined that the material capital investment, technological level and industrial structure, as important pushing factors in economic growth, have more obvious characteristics of diversification in late developing Large Countries,[4] and it is these characteristics of diversification that call for adaptation of heterogeneous human capital to it to allow their role to be brought into play. It is true that the human capital in late-developing Large Countries has very obvious heterogeneity characteristics, as demonstrated by the fact that in some regions there are fairly rich professionals on high and new technologies, while in other regions, there are abundant ordinary professionals with practical operation skills. From this, it can be seen that the economy of late-developing Large Countries is a complicated system incorporating material capital investment, technological level, industrial structure as well as human capital with diversification characteristics. Its growth depends on the synergistic effect of human capital with all subsystems of industrial structure, technological level and material capital investment, and such synergistic effect can be described by the term "coupling" in physics.

Coupling is a phenomenon of two or more systems or forms of movement influencing with each other to reach synergy by way of various interactions, and is a dynamic association of mutual dependence, coordination and promotion between all subsystems in favorable interaction. The key for a system to change from an orderless to an orderly mechanism lies in the synergy between the parameters within the system, which governs the relevant characteristics and laws of the system, and the coupling degree is just a measurement of this synergy. It is in view of the important effect of the dynamic relations of mutual dependence, coordination and promotion between the human capital system incorporating both quantity and quality factors on one hand and the systems of industrial structure, technological level and material capital investment on the other, that this article applies the coupling theory to describe the extent of influence between the human capital system and the systems of industrial structure, technological level and material capital investment, and to reveal the human capital comprehensive advantage of late-developing Large Countries and its functioning mechanism.[5] This analysis framework shows that, the dynamic coupling relations between the human capital system and the systems of industrial structure, technological level and material capital investment are in essence the objective manifestation of the mutual matching and depending relations between these systems and between the factors within these systems. If mutual influence and even synergy can be reached via various interactions with the subsystems of industrial structure, technological level and material capital investment, i.e. the coupling action is strong, even human capital at fairly low level can also promote rapid economic growth.

2. Connotation of human capital comprehensive advantages in late-developing Large Countries

By adopting the analysis framework of system theory and coupling theory, and with full consideration of the objective fact that the overall human capital level is low with serious shortage in late-developing Large Countries, the author defines the human capital comprehensive advantages in late-developing Large Countries as: a comparative advantage formed by fairly strong coupling effect of the heterogeneous human capital with the diversified material capital investment, industrial structure and technological level in late-developing Large Countries under the realistic condition that their human capital level is far below that in developed countries. This comprehensive advantage can be measured with degree of coupling, the higher degree of coupling indicates higher degree of matching between the human capital and the material capital investment, industrial structure and technological level, therefore more apparent comprehensive advantages, and more powerful promoting role of human capital in economic growth.[6]

3. Functioning mechanism of human capital comprehensive advantages in late-developing Large Countries

In developing countries, although the accumulation of material capital is a main drive factor in economic growth, the role of technological progress is daily enhancing, and the optimization and upgrading of industrial structure also play a very prominent role. In the meantime, human capital has a crucial effect on the roles of the three.[7] In the objective reality that the overall human capital level is low with serious shortage, late-developing Large Countries have a rational choice to bring into full play the promoting role of existing human capital in economic growth by enhancing the matching efficiency of human capital with material capital investment, technological level and industrial structure following the comparative advantage. Therefore, the mechanism of human capital comprehensive advantages promoting economic growth in late-developing Large Countries can be expressed with Fig. 7.1.

According to Fig. 7.1, the mechanism of human capital comprehensive advantages promoting economic growth in late-developing Large Countries can be expounded as follows:

First, the dynamic coupling of heterogeneous human capital and diversified material capital investment can greatly improve the application efficiency of material capital investment, thereby pushing forward the rapid development of economy. The effects of material capital and human capital are complementary, although the accumulation of material capital is the main drive force in the economic growth in developing countries, it will be difficult to bring into full play the role of material capital if the material capital investment is increased on one side without being matched with by corresponding human capital. Late-developing Large Countries have diversified types of material capital investment (for example, there are state-owned investment, collective investment, private investment and foreign direct investment, etc.), and with great difference between regions. The matching of heterogeneous human capital with diversified material capital investment can on one hand allow the existing material capital to give play to maximum productive force, and on the other hand promote further accumulation of material capital, in turn, the further accumulation of material capital requires human capital

异质性人力资源 Heterogeneous human capital
多元化物质资本投资 Diversified material capital investment
多元化技术水平 Diversified technological level
多元化产业结构 Diversified industrial structure
经济增长 Economic growth 匹配 Matching

Fig. 7.1 Functioning mechanism of human capital comprehensive advantages in late-developing Large Countries

at higher levels to match with it, so this dynamic adjustment can become an important impetus to the rapid economic growth.

Second, the dynamic coupling of heterogeneous human capital with diversified technological level enables late-developing Large Countries to foster competitive advantage while making use of their current later-mover advantage, so that late-developing Large Countries can realize leap-forward development. Late-developing Large Countries have both adaptive technologies and high and new technologies, and conduct technological innovation while importing and imitating advanced technologies of developed countries; corresponding to this technical foundation are comparative advantage industries mainly based on adaptive technologies or potential comparative advantage industries, and industries with competitive advantage mainly supported by high and new technologies. The matching of heterogeneous human capital with diversified technological levels in late-developing Large Countries enables effective innovation, absorption and utilization of all types of technologies, thereby boosting the rapid development of industries founded on corresponding technologies. Specifically, some types of human capital in late-developing Large Countries can match with existing comparative advantage industries mainly adopting adaptive technologies, so as to ensure the full play of existing comparative advantage; some types of human capital can adapt with potential comparative advantage industries demanding higher technological levels, thereby promoting the potential comparative advantage to convert to realistic comparative advantage, and in the meantime, some other types of human capital (mainly innovation-oriented human capital with high technological proficiency) can effectively match with high-tech industries and emerging industries founded on high and new technologies, thereby creating early-mover advantages and fostering competitive advantage. The development of early-mover advantages and competitive advantage depends on corresponding technological foundation, and changes of technological foundation in turn can stimulate the corresponding adjustment of human capital structure, this self-enhancing relation between the two can help late-developing Large Countries realize economic catch-up by implementing a leap-forward development strategy.

Third, the dynamic coupling of heterogeneous human capital with diversified industrial structure is conducive to speeding up the optimization and upgrading of industrial structure, to realize rapid development of economy. The optimization and upgrading of industrial structure also play a quite prominent role in promoting the economy of developing countries. The industrial structure of late developing Large Countries is characterized by diversity, with labor intensive industries and capital intensive industries, as well as technology intensive industries, and there are also quite big differences between regions in the development level and distribution of industries. Late-developing Large Countries can effectively avoid the "comparative interest trap" and ensure the sustained rapid growth of economy only by promptly upgrading their industrial structure, i.e. promptly transforming from the labor intensive industries to capital and technology intensive industries, and from the previous later-mover advantage of low labor cost and technical imitation to the later-mover advantage relying on human capital investment and technological innovation. The coupling of heterogeneous human capital with the diversified industrial structure can provide a powerful support to the optimization and upgrading of industrial structure. Specifically, with their fairly high economic development level and relatively rich material capital, developed regions will focus on developing capital intensive industries at the leading edge of new technologies and high and new technologies to drive the optimization and upgrading of their industrial structure, and gradually transfer the labor intensive industries to underdeveloped regions. The relatively rich high-tech professionals in developed regions will be the important drive force for them to develop high-tech industries, while the relatively abundant ordinary professionals in underdeveloped regions can serve as a solid foundation for them to accept the industrial transfer from developed regions. With the effective matching of corresponding types of human capital and their positive guiding, the industrial structure in late-developing Large Countries can be adjusted and upgraded quickly. The optimization of industrial structure can in turn enhance the stimulation of human capital investment, therefore the favorable interaction between the two can help realize rapid economic growth.

The above analysis shows that, there are dynamic coupling relations of mutual influence and promotion between the heterogeneous human capital on one hand and diversified material capital investment, technological level and industrial structure on the other, and such dynamic coupling relations enable self-enhancing between the human capital structure adjustment and material capital accumulation, technological level upgrading and industrial structure optimization, thereby pushing the sustained rapid growth of economy.

7.3 Empirical Verification with Data of China

On the basis of the connotation and functioning mechanism of human capital comprehensive advantages in late-developing Large Countries, we put forward the following basic hypothesis: there is fairly high degree of coupling or matching between the heterogeneous human capital and diversified industrial structure, material capital investment and technological level in late-developing Large Countries, and it is this effective matching that enables the human capital at comparatively low level to powerfully promote the economic growth. In the following, we will make empirical verification of this hypothesis.

1. Building the coupling degree model

On the basis of the analysis framework of this book, and taking into account the complexity and association of heterogeneous human capital with industrial structure, material capital investment and technological level system, we will use the coupling degree model (Liu Sifeng et al. 2004; Liu Yaobin et al. 2005; Bi Qige et al. 2007) to obtain the coupling degree of human capital and industrial structure system, human capital and material capital investment system, and human capital and technological level system, to describe the coordination extent of coupling between them, and analyze the role of human capital in economic growth. The basic steps are as follows:

(1) Building the analysis systems and indicator systems.

The analysis systems in this book include the "human capital—industrial structure" system, "human capital—material capital investment" system, and "human capital—technological level" system, therefore, we determined the relevant analysis series and their indicator systems as follows:

① Human capital series group (X_i).

The empirical research in this section is aimed at inspecting the matching degree of heterogeneous human capital with industrial structure, material capital investment and technological level, therefore we set the indicator system of the human capital series group: population with primary school education level (X_1), population with junior secondary school education level (X_2), population with senior secondary school education level (X_3), population with secondary professional education level (X_4), population with junior college education level (X_5), population with undergraduate education level (X_6), population with post-graduate education level (X_7), per student educational fund in primary schools (X_8), per student educational fund in junior secondary schools (X_9), per student educational fund in senior secondary schools (X_{10}), per student educational fund in higher schools (X_{11}), student to teacher ratio in primary schools (X_{12}), student to teacher ratio in regular secondary schools (X_{13}), student to teacher ratio in secondary professional schools (X_{14}), and student to teacher ratio in higher schools (X_{15}).[8]

② Industrial structure series group (Y_i)

According to the GDP composition data in *China Statistical Yearbook*, we set the indicator system for the industrial structure series group as: first industry (Y_1), secondary industry (Y_2), building industry (Y_3), agriculture, forestry, animal husbandry and fishery service industry (Y_4), geological survey and water conservancy administration (Y_5), communications, transport, warehousing, post and telecommunication (Y_6), whole-sale and retail trade and catering industry (Y_7), banking and insurance industry (Y_8), real estates (Y_9), social service industry (Y_{10}), public health, sports and social welfare (Y_{11}), education, culture, art and broadcasting and television (Y_{12}), scientific research and comprehensive technology service undertakings (Y_{13}), state organs, Party and government organs and social groups (Y_{14}), and other industries (Y_{15}).

③ Material capital investment series group (Z_i).

In building the material capital investment series group, we use the indicators of fixed asset investment grouped by economic category: state-owned economy (Z_1), collective economy (Z_2), individual economy (Z_3), joint venture economy (Z_4), stock-system economy (Z_5), foreign invested economy (Z_6), Hong Kong, Macau and Taiwan invested economy (Z_7), and other economy (Z_8).

④ Technological level series group (P_i),

Indicators to measure technological progress include: quantity of domestic technological invention patents, quantity of imported technologies and equipment, and amount of FDI, as it is difficult to obtain data on imported technologies and equipment, our technological level series group only includes the following: quantity of invention patents approved (P_1), quantity of utility model patents approved (P_2), quantity of appearance design patents approved (P_3) and FDI (P_4).[9]

(2) Making the relevant data dimensionless.

As the original data of indicators have different dimensions, we followed the practice by Liu Yaobin (2005) and Bi Qige et al. (2007), to render the indicator data in all analysis series groups dimensionless using the maximum difference normalization method:

$$Z_{ij} = \frac{X_{ij} - \min_i X_{ij}}{\max_i X_{ij} - \min_i X_{ij}} \tag{7.1}$$

where X_{ij} is the original values of the indicators in relevant analysis series groups.

(3) Obtain the gray relational coefficient.

$$\varepsilon_{ij}(t) = \frac{\min_i \min_j \left| z_i^X(t) - z_j^Y(t) \right| + \rho \max_I \max_J \left| Z_i^Y - Z_j^Y \right|}{\left| Z_i^X - Z_j^Y \right| + \max_I \max_J \left| Z_i^X - Z_j^X \right|} \tag{7.2}$$

where $Z_i^X(t)$, $Z_j^Y(t)$ are respectively the standardized values of the relevant indicators in the analysis series groups, ρ the resolution ratio, normally taken as 0.5, and $\xi_{ij}(t)$ the correlation coefficient at the time t.

(4) Calculate the relational degree and coupling degree.

To calculate the average value of correlation coefficient with respect to the number of samples, a relational degree matrix γ can be obtained, which reflects the complicated relations of coupling function among the objects in the system to be analyzed. The calculation formula for relational degree is:

$$\gamma_{ij} = \frac{1}{n} \sum_{j=1}^{n} \varepsilon_{ij}(t) \quad (n = 1, 2, \ldots) \tag{7.3}$$

In formula (7.3), n is the number of samples, or the number of indicators for human capital (or industrial structure, material capital investment or technological level) selected in this section. The extent of close relations between the indicators in the systems can be analyzed by comparing the magnitude of relational degree γ_{ij}. It is low relational degree when $0 < \gamma_{ij} \leq 0.35$, with weak coupling effect between the two system indicators; it is medium relational degree when $0.35 < \gamma_{ij} \leq 0.65$, with medium coupling effect between the two system indicators; it is fairly high relational degree when $0.65 < \gamma_{ij} \leq 0.83$, with fairly strong coupling effect between the two system indicators; and if $0.85 < \gamma_{ij} \leq 1$, the two system indicators change almost consistently with each other, and there is an extremely strong coupling effect between them.

On the basis of the relational degree matrix, average values of γ_{ij} are calculated by line or row, to obtain the average relational degree of a given indicator in an analysis series group with another analysis series group:

$$D_i = \frac{1}{l}\sum_{j=1}^{l}\gamma_{ij} \quad (i = 1, 2, \ldots, m;\ j = 1, 2, \ldots, l) \tag{7.4}$$

$$D_j = \frac{1}{m}\sum_{i=1}^{m}\gamma_{ij} \quad (i = 1, 2, \ldots, m;\ j = 1, 2, \ldots, l) \tag{7.5}$$

The above-mentioned average relational degree can be used to determine the most important factor in the mutual influence of systems. In addition, to make overall analysis of the coordination extent of system coupling, we use the following formula to calculate and analyze the coupling degree of systems:

$$C(t) = \frac{1}{m \times l}\sum_{i=1}^{m}\sum_{j=1}^{l}\varepsilon_{ij}(t) \tag{7.6}$$

where, C(t) is the coupling degree, and m and l are respectively the number of indicators of relevant analysis series group.

2. Selection of samples and sources of data

In this section, only 31 provinces, autonomous regions and municipalities of China in 2000 were taken as the analysis samples, because: the fairly accurate number of population of different educational status in the human capital subsystem can be obtained from the data of the third (1982), fourth (1990) and fifth (2000) population census, however, only in the fifth census the education status is classified into seven categories of primary school, junior secondary school, senior secondary school, secondary profession school, junior college, undergraduate, post-graduate; (2) even if the education status is classified into four categories of primary school, junior secondary school, senior secondary school (including secondary professional), junior college and above (including undergraduate and post-graduate), so that the statistical caliber of number of population of different education status in the above three census is identical, the statistical caliber of some other indicators such as industrial structure and material capital investment is not identical for different years. The relevant data required for building the indicator systems of analysis systems were taken from *China Statistical Yearbook* (2001), *China Demographic Yearbook* (2001), *China Educational Finance Statistical Yearbook* (2001).

3. Analysis of results

With the afore-said coupling degree model, we first used the whole country as a sample, to make empirical verification of the coupling degree of "human capital—industrial structure" system, "human capital—material capital investment" system, "human capital—technological progress" system (the results are respectively presented in Tables 7.1, 7.2 and 7.3), to explore on the dynamic correlations of mutual dependence, mutual coordination and mutual promotion of human capital with industrial structure, material capital investment and technological level, and to analyze on this basis the important role of human capital in economic growth.

(1) Coupling relations of human capital—industrial structure system

It can be seen from Table 7.1 that, the total coupling relational degree of human capital with industrial structure is 0.6737, indicating a fairly strong interactive coupling between them, that is, the heterogeneous human capital and diversified industrial structure have fairly high matching degree. In detail, all indicators of human capital have an average relational degree of over 0.57 with industrial structure, indicating their important influence on the industrial structure. Among them, the top 5 indicators with the highest relational degree with industrial structure are population with post-graduate education level (0.7482), per student educational fund in senior secondary schools (0.7262), population with undergraduate education level (0.7154), per student educational fund in primary schools (0.7093), per student educational fund in junior secondary schools (0.6977), indicating these are the main factors acting upon the industrial structure; the average relational degree of quantity indicators (the number of population of different education level) and quality indicators (per student educational fund and student to teacher ratio) of human capital with industrial structure are respectively as high as 0.6839 and 0.6648, indicating that both of them are important aspects governing the role of the human capital, therefore, in studying the role of human capital, both quantity and quality should be taken into consideration.[10] In the quantity of human capital, the relational degree of population with post-graduate, undergraduate education level with industrial structure is at a high level, the relational degree of human capital with junior college education level with industrial structure (0.6351) is at medium level, and the coupling of population with other education level is also fairly strong (the average relational degree of population with primary school, junior secondary school, senior secondary school and secondary professional education levels with the industrial structure is respectively 0.6811, 0.6941, 0.6571 and 0.6562 in that order), it is this pattern that the population with both high and low educational levels have fairly high relational degree with industrial structure that enables the human capital status of China with a small number of population at high

Table 7.1 Coupling degree and relational degree of human capital—industrial structure system

Indicator	X_1	X_2	X_3	X_4	X_5	X_6	X_7	X_8	X_9	X_{10}	X_{11}	X_{12}	X_{13}	X_{14}	X_{15}	Average
Y_1	0.6657	0.6699	0.6290	0.6702	0.6973	0.7746	0.6792	0.6763	0.6529	0.6708	0.7030	0.7118	0.6510	0.6874	0.5831	0.6748
Y_2	0.6850	0.6906	0.6590	0.5911	0.6056	0.7661	0.7550	0.7082	0.6792	0.7216	0.7207	0.6372	0.6554	0.6248	0.6156	0.6743
Y_3	0.6901	0.7074	0.6747	0.6994	0.6680	0.7889	0.6744	0.6709	0.6463	0.6854	0.7149	0.6700	0.6815	0.6679	0.6108	0.6834
Y_4	0.6788	0.7237	0.6783	0.6485	0.6630	0.7352	0.8477	0.7795	0.7487	0.7875	0.7817	0.6059	0.6483	0.6393	0.5117	0.6985
Y_5	0.6958	0.6475	0.6399	0.6628	0.6792	0.7569	0.7279	0.7224	0.6978	0.7747	0.0183	0.6917	0.7034	0.6235	0.6294	0.6448
Y_6	0.7051	0.6821	0.6504	0.6464	0.5969	0.7289	0.7740	0.7021	0.6942	0.7209	0.7629	0.6730	0.6631	0.6137	0.5550	0.6779
Y_7	0.6675	0.7097	0.7067	0.6711	0.6492	0.7690	0.7226	0.6923	0.6502	0.6827	0.6953	0.6331	0.6099	0.6527	0.6031	0.6743
Y_8	0.6853	0.6819	0.6372	0.6334	0.6095	0.6178	0.7852	0.7128	0.7154	0.7791	0.7097	0.6596	0.6233	0.5479	0.5227	0.6614
Y_9	0.664I	0.6684	0.6266	0.6148	0.6025	0.6839	0.8439	0.7349	0.7459	0.7526	0.7592	0.6303	0.6228	0.5656	0.5302	0.6697
Y_{10}	0.6567	0.6964	0.6711	0.6770	0.6536	0.6444	0.7960	0.7274	0.7015	0.7697	0.7426	0.6457	0.6216	0.5980	0.5567	0.6778
Y_{11}	0.6613	0.6688	0.5873	0.6902	0.5932	0.7620	0.6921	0.6907	0.6571	0.6645	0.6438	0.6462	0.6104	0.6866	0.6262	0.6587
Y_{12}	0.7171	0.7434	0.7015	0.7210	0.6493	0.6616	0.6666	0.6789	0.6768	0.6677	0.6689	0.6350	0.6231	0.7452	0.6206	0.67&5
Y_{13}	0.6533	0.6617	0.6287	0.6348	0.6037	0.6138	0.7413	0.6849	0.7606	0.7981	0.6744	0.5999	0.5975	0.6389	0.5260	0.6545
Y_{14}	0.7091	0.7666	0.7191	0.6513	0.6404	0.7751	0.6956	0.6689	0.6745	0.6755	0.6751	0.6625	0.7110	0.6792	0.6050	0.6873
Y_{15}	0.6808	0.6941	0.6397	0.6311	0.6144	0.6527	0.8221	0.7898	0.7645	0.7429	0.9831	0.6213	0.6102	0.5902	0.5147	0.6901
Average	0.6811	0.6911	0.6571	0.6562	0.6351	0.7154	0.7482	0.7093	0.6977	0.7262	0.6836	0.6482	0.6422	0.6374	0.5740	
	0.6839							0.6648								
	Coupling degree of human capital—industrial structure: 0.6737															

Source Calculated and obtained by the author

Table 7.2 Coupling degree and relational degree of human capital—material capital investment system

Indicator	X_1	X_2	X_3	X_4	X_5	X_6	X_7	X_8	X_9	X_{10}	X_{11}	X_{12}	X_{13}	X_{14}	X_{15}	Average
Z_1	0.7079	0.7276	0.7044	0.6748	0.6418	0.7377	0.6548	0.6637	0.6482	0.6700	0.6348	0.6340	0.6628	0.6699	0.6249	0.6705
Z_2	0.6243	0.6757	0.6257	0.5812	0.5996	0.6842	0.8320	0.7406	0.7317	0.7447	0.7250	0.6189	0.6280	0.6105	0.5659	0.6661
Z_3	0.7186	0.7302	0.7061	0.7194	0.7008	0.7806	0.7005	0.6621	0.6740	0.7113	0.7090	0.6861	0.6417	0.6729	0.5828	0.6931
Z_4	0.6975	0.7103	0.6797	0.6522	0.6519	0.6890	0.8221	0.7159	0.7443	0.7360	0.7469	0.6360	0.6410	0.6582	0.5574	0.6892
Z_5	0.6262	0.6831	0.6815	0.6218	0.6656	0.6441	0.7764	0.7653	0.7437	0.7700	0.7240	0.6127	0.6307	0.6447	0.5737	0.6776
Z_6	0.6412	0.6596	0.6406	0.6166	0.6135	0.6236	0.8023	0.7767	0.7615	0.7737	0.7030	0.6804	0.5905	0.5743	0.5382	0.6664
Z_7	0.6454	0.6676	0.6373	0.6544	0.6186	0.5893	0.8044	0.8005	0.7734	0.7276	0.6486	0.6663	0.5844	0.5641	0.5100	0.5595
Z_8	0.678J	0.6987	0.7176	0.7037	0.6746	0.7232	0.7829	0.7376	0.7500	0.7464	0.7552	0.6193	0.6012	0.6604	0.5743	0.6949
Average	0.6674	0.6941	0.6741	0.6530	0.6458	0.6840	0.7719	0.7328	0.7287	0.7350	0.7058	0.6442	0.6225	0.6319	0.5659	
	0.6843							0.6709								
	Coupling degree of human capital—material capital investment: 0.677 1															

Source Calculated and obtained by the author

Table 7.3 Coupling degree and relational degree of human capital—technological level system

Indicator	X_1	X_2	X_3	X_4	X_5	X_6	X_7	X_8	X_9	X_{10}	X_{11}	X_{12}	X_{13}	X_{14}	X_{15}	Average
P_1	0.6561	0.6802	0.6648	0.6477	0.6075	0.6068	0.7499	0.7836	0.7807	0.4950	0.6828	0.6112	0.6665	0.6641	0.5882	0.6594
P_2	0.6473	0.6970	0.6569	0.6341	0.6424	0.6201	0.7548	0.7298	0.7130	0.7318	0.6949	0.6229	0.6412	0.7156	0.5786	0.6720
P_3	0.6254	0.6810	0.6447	0.6221	0.6039	0.6233	0.8227	0.8291	0.8025	0.7520	0.7113	0.6561	0.6266	0.6302	0.5209	0.6768
P_4	0.6463	0.6756	0.6575	0.6262	0.6092	0.6201	0.8239	0.7718	0.7796	0.7429	0.6862	0.6240	0.6343	0.6164	0.5310	0.6699
Average	0.6438	0.6835	0.6560	0.6325	0.6157	0.6176	0.7878	0.7786	0.7703	0.6804	0.6938	0.6285	0.6421	0.6566	0.5554	
	0.6624							0.6757								
	Coupling degree of human capital—technological level: 0.6995															

Source Calculated and obtained by the author

educational level and large number of population at low educational level to fairly well match with the industrial structure.[11] In the quality indicators of human capital, the average relational degree of per student educational fund in primary schools, junior secondary schools and senior secondary schools with industrial structure is not only higher than other quality indicators, but also higher than the average relational degree of population with primary, junior secondary and senior secondary education levels with industrial structure, reflecting the relative importance of education input in primary, junior secondary and senior secondary schools.[12] It should be pointed out that, the average relational degree of all indicators of industrial structure with human capital is over 0.64, indicating that the industrial structure also has non-ignorable effect on and role in human capital investment.[13]

(2) Coupling relations of human capital—material capital investment system

Table 7.2 shows that, the total coupling degree of human capital with material capital investment is 0.6771, indicating fairly strong interactive coupling between them, i.e. they have fairly high matching degree. The population with post-graduate education level has the most obvious effect on material capital investment (relational degree is as high as 0.7719), the per student educational fund in primary, junior secondary and senior secondary schools, and in higher schools also has important effect on material capital investment, all with average relational degree of over 0.7 with material capital investment, and the quantity and quality of human capital are also quite important to the application efficiency of material capital investment (the average relational degree of quantity and quality indicators with material capital investment is respectively 0.6843 and 0.6709). It can be known from the average relational degree of the quantity of human capital with material capital investment that, population with both high and low educational levels has fairly strong coupling effect with material capital investment, so this can also ensure the adaptability of the human capital status of China with a small number of population at high educational level and large number of population at low educational level to the diversified material capital investment.

(3) Coupling relations of human capital—technological level system

It can be known from Table 7.3 that, the total coupling degree of human capital with technological level is 0.6995, indicating a high matching degree between the two. Population with post-graduate education level has the highest average relational degree with technological level, indicating its highest influence on the technological level, other human capital indicators with average relational degree over 0.7 include per student educational fund in primary schools (0.7786) and per student educational fund in junior secondary schools (0.7705), the average relational degree of human capital quantity and quality indicators with technological level is respectively 0.6624 and 0.6757, indicating an important role of the quantity and quality of human capital in upgrading the technological level. Analysis of the relation of human capital quantity indicators with technological level also shows a fairly strong correlation between population with low educational level (population with junior secondary and senior secondary education level) and technological level, therefore, the human capital status of China with a small number of population at high educational level and large number of population at low educational level can also well match with the diversified technological level.

Then, we made further empirical research on the coupling relations of human capital with industrial structure, material capital investment and technological development level with the east, central and west regions as samples, and the result is as shown in Table 7.4.

It can be known from Table 7.4 that, in all regions, the quantity indicator of human capital has fairly high relational degree with industrial structure, material capital investment and technological level, indicating that population of different educational levels all play an important role in them. The average relational degree of the quality indicator of human capital with industrial structure, material capital investment and technological level is also strong, also indicating the importance of the quality of human capital. As a whole, in all regions, the coupling degree of human capital with industrial structure, material capital investment and technological level is over 0.6, further proving a fairly strong interactive coupling between them, or a fairly high matching degree.

In summary, the empirical research of the whole country and by regions show that, there is a fairly strong dynamic coupling relationship between the heterogeneous human capital on one hand and the diversified industrial structure, material capital investment and technological level on the other. It is this dynamic coupling that gives comprehensive advantages to the human capital in late-developing Large Countries, and with such comprehensive advantages, even the human capital at comparatively low level in late-developing Large Countries can tremendously promote the economic growth.

7.3.1 Conclusions and Policy Recommendations

In this section, the author has tried to study the theoretical hypothesis "although the human capital level in late-developing Large Countries is low, it still can promote fairly rapid economic growth", based on the realistic characteristics of late-developing Large Countries—regional diversification, economic diversification and technological diversification, and in the view angle of the heterogeneity of human capital and its adaptability to the diversified industrial structure, material capital investment and technological development level. Our theoretical analysis shows that, because the human capital can well match with diversified industrial structure, material capital investment and technological level, the human capital level at fairly low level in late-developing Large Countries can still tremendously promote economic growth. The empirical research in this chapter supports this theoretical hypothesis, with the basic conclusions that: both quantity and quality of human capital have important effect on the industrial structure, material capital investment and technological level, and further to economic growth, therefore, in

Table 7.4 Coupling degree and relational degree of human capital with industrial structure, material capital investment and technological level in three regions

		Industrial structure			Material capital investment			Technological level		
		East	Central	West	East	Central	West	East	Central	West
Human capital	Quantity indicator	0.6536	0.6613	0.6973	0.6581	0.6534	0.6839	0.6626	0.6472	0.6607
	Quality indicator	0.6501	0.6614	0.6704	0.6456	0.6530	0.6653	0.6519	0.6517	0.6524
	Coupling degree	0.6517	0.6614	0.6829	0.6514	0.6532	0.6639	0.6569	0.6496	0.6563

The east region includes Beijing, Tianjin, Hebei, Shanghai, Jiangsu, Zhejiang, Fujian. Shandong, Guangdong, Hainan and Liaoning; the central region includes Shanxi, Heilongjiang, Jilin, Anhui, Jiangxi, Henan, Hubei and Hunan; and the west region includes Inner Mongolia, Guangxi, Sichuan, Chongqing, Guizhou, Yunnan, Tibet, Shaanxi, Gansu, Qinghai, Ningxia and Xinjiang

Source Calculated and obtained by the author

analyzing the roles of human capital, it is necessary to take it as a system incorporating both quantity and quality; the population with post-graduate education level has the highest relational degree with industrial structure, material capital investment and technological level; relative to the indicators of population with primary school, junior secondary school and senior secondary school education levels and other quality indicators, educational input for primary schools, junior secondary schools and senior secondary school produce more effect on the industrial structure, material capital investment and technological level; both population with high education level and population with low education level have fairly high relational degree with industrial structure, material capital investment and technological level, therefore, the human capital status of China with a small number of population at high educational level and large number of population at low educational level can also well match with the diversified industrial structure, material capital investment and technological level, as demonstrated by the fairly high coupling and fairly strong dynamic coupling relations of the heterogeneous human capital with industrial structure, material capital investment and technological level. It is this fairly strong interactive coupling between these systems that the fairly low human capital level in late-developing Large Countries can promote the sustained rapid economic growth.

On the basis of the empirical conclusions in this chapter, we come to the opinion that developing Large Countries, under the realistic conditions of fairly low level human capital, should pay more attention to enhancing the coupling or matching degree of their human capital with the diversified industrial structure, material capital investment and technological level, only by this can they avoid their shortcomings and bring into full play the role of existing human capital, to push forward the sustained, rapid and coordinated development of economy, specifically:

(1) When increasing human capital investment, developing Large Countries should emphasize on increasing the proportion of high quality human capital, and make more investment in primary and secondary education, and this is an important route to enhance the coupling degree of human capital with industrial structure, material capital investment and technological level. Currently in developing countries, the population with high educational level is small while that with low educational level is large, and the investment in primary and secondary education is obviously insufficient. China should pay attention to these two issues in its human capital investment. Our research has revealed that the population with post-graduate educational level and the per student educational fund in primary schools, junior secondary and senior secondary schools are highly correlated with the industrial structure, material capital investment and technological level. Therefore, on one hand, we should increase the number of population with high educational level and raise the proportion of high-quality human capital by strengthening the post-graduate education; on the other hand, we must ensure the steady development and upgrading of primary and secondary education by increasing the investment in these areas. This will help raise the overall level of human capital, and enhance

the coupling degree of human capital with industrial structure, material capital investment and technological level.

(2) Developing Large Countries should energetically increase the allocation efficiency of human capital, and this is the fundamental way to increase the coupling degree of human capital with industrial structure, material capital investment and technological level, and form and give play to the human capital comprehensive advantages. At the macroscopic level, today, the proportion of high educational level population in the east region is higher than that in the central and western regions, while the proportion of low educational level population in the latter is higher than that of the former. In this sense, the development strategy that the central and western regions take initiative to accept the transfer of labor intensive industries from the east region, and the east region energetically develop the capital intensive industries at the leading edge of new technologies and high and new technology industries can help raise the human capital allocation efficiency. Therefore, governments in the central and western regions should actively foster and develop competitive, open and well-organized markets, increase the efforts in institutional innovation, and adopt appropriate preferential policies on taxation and finance, to guide in suitable industries and enterprises; in the meantime, they should increase working efficiency of government and enhance the concept of credibility, to create a good environment favoring the acceptance of industrial transfer. At the microscopic level, the allocation efficiency of human capital depends on whether "everyone is given full scope to their talent", and professionals of different levels and types should be arranged at suitable posts so that they can well play their roles. Therefore, governments should establish complete social security systems, actively foster regional open human capital market, push forward the reform of household registration system, break the "department ownership" of professionals, speed up the process of urbanization, and implement overall planning for urban and rural areas and the fostering programs for all types of professionals, to guide the rational circulation of human capital. In general, a developing Large Country can realize high degree of coupling of heterogeneous human capital with diversified industrial structure, material capital investment and technological level only by adopting effective measures on both macroscopic and microscopic aspects on its actual national conditions to increase the allocation efficiency of human capital, so that the present low level human capital can best play its role in promoting economic growth.

The research in this section certainly has some shortcomings, for example, due to the restriction of available data, no empirical analysis could be made on the dynamic changes of comprehensive advantages; and comparative research with other developed countries is also lacking, therefore limiting the convincing power of the research conclusions. However, these shortcomings are what we will work on further to make improvement.

Notes

1. Many scholars have analyzed the close link of human capital and economic growth, and made empirical inspection with data of China. Refer to: Cai Fang and Du Yang: *Convergence and Differences in Regional Economic Growth in China—Enlightenment to the Strategy of Developing the West*, *Economic Research* 2000 Vol. 10; Shen Kunrong, Ma Jun: *The "Club Convergence" Characteristic in Economic Growth in China and Analysis of Its Origin*, *Economic Research* 2002 Vol. 1.
2. The statements by Justin Yifu Lin and Li Yongsheng (2003) that enterprises in countries with high material capital level must employ professionals with high human capital; the question from Zou Wei and Dai Qian (2003) on matching of technology import with human capital; the research by Guo Jiqiang (2005) on the matching of human capital investment structure and industrial structure, all imply the importance of heterogeneous human capital to the adaptability of material capital investment, technological level and industrial structure.
3. Yao Ouyang: *The Direct Outbound Investment Strategy of China Based on Comprehensive Advantages Of Large Country*, *Economics of Finance and Trade* 2006 Vol. 5.
4. The diversified characteristics of material capital investment and technological level, industrial structure in late developing Large Countries are demonstrated not only by their diversified types, but also by the great differences between regions.
5. Although coupling is a concept in physics, it has now been extensively applied in the researches in social science fields. For example, Liu Yaobin et al. (2005) used a coupling degree model to reveal the coordinated development relations between regional urbanization and ecological environment in China; Bi Qige et al. (2007) used the coupling degree model to study the main factors of coupling between the population structure and regional economy in Inner Mongolia and coupling relations.
6. It is worth pointing out that, the purpose of this section to discuss the connotation of human capital comprehensive advantages in late developing Large Countries is to describe the role of human capital at fairly low level in late developing Large Countries in promoting economic growth, and it does not deny the human capital advantages in developed countries or regions. As compared with developing countries, developed countries or regions have obvious advantages of human capital itself in both quantity and quality.
7. Justin Yifu Lin and Li Yongjun (2003) pointed out that, only with input of material capital, without the corresponding upgrading of human capital and its cooperation, the newly invested machinery and equipment cannot play their maximum productive force. Zou Wei and Dai Qian (2003), based on a standard endogenous growth model, analyzed the important role of human capital level in enhancing the technological imitation and absorption ability of a country. The research by Dai Qian and Bie Zhaoxia (2000) using a dynamic comparative advantage model shows that, the increase of human capital level is critical to the upgrading of industrial structure in developing countries.

8. The "Father of human capital" Schultz pointed out that, human capital includes two aspects of "quantity" and "quality", and he emphasized more on the latter. The per-student educational fund and student to teacher ratio can be taken as substituting indicators for human capital (education) quality (Rarro and Lee 1996).
9. FDT catches technological transfer from foreign countries, and patent data reflects the technological progress resulted from domestic research and development activities, and patents of different types have differences in technological level.
10. Their high relational degree with material capital investment and technological level also indicates this point. In fact, the higher human capital quality, the stronger their ability in resources allocation and absorbing technologies, hence more prominent roles.
11. In 2000, the proportion of population with primary school and junior secondary school education level was respectively as high as 38.18 and 36.52 %, while the proportion of population with senior secondary school, secondary professional, junior college, undergraduate and post-graduate education level was only respectively 8.56, 3.39, 2.51, 1.22 and 0.08 %.
12. The average relational degree of per student educational fund of primary schools, junior secondary schools, senior secondary schools with material capital investment and technological level has also proved this point.
13. The high relational degree of material capital investment and technological level indicators with human capital also indicates their important effect on human capital.

Chapter 8
Case: Regionally Coordinated Development of Late-Developing Large Countries

In this chapter, regional development is taken as a case to make detailed analysis of regional coordination advantages in late developing Large Countries. The technology adaptive capacity in late-developing regions of late-developing Large Countries is not an exogenous variable, instead, it is endogenous in the selection of suitable technologies in late-developing regions. Economic convergence can be realized rapidly if the technologies introduced in late-developing regions are adaptive to the existing technological foundation, human capital and material capital. The technology adaptive capacity on the basis of technologies diversification was analyzed by adding the adaptive capacity parameter in the expanded "leader–follower" model, and it was found that the coordinated development of regional economy in late-developing Large Countries is determined by the extent of technological diversification and human capital gap of the two regions and the advanced level of the human capital and the imported technologies in the late-developing region; adaptation of the diversified technologies in late-developing regions with the human capital, material capital, industrial structure as well as regional economy development level within the region can promote the coordinated development of regional economy; however, only with the conditions of technological diversification and adaptability, late-developing regions cannot converge to the same per capita output and wage level as in developed regions. On the basis of this conclusion, the author analyzed the mechanism of coordinated development of regional economy in late-developing Large Countries, and proposed the corresponding policy recommendations.

8.1 Related Literatures

Conditions in developing countries differ greatly, and there are emerging developing countries like Korea and Singapore with small national territorial area, simple resources, limited population size and simple development level, as well as

Y. Ouyang, *The Development of BRIC and the Large Country Advantage*,
DOI 10.1007/978-981-10-0633-3_8

late-developing Large Countries like China and India with large national territorial area, rich resources, large population, and diversified development levels.[1] Therefore, there is great restriction in the traditional development economics in analyzing late-developing Large Countries, that is, the diversified reality of late-developing Large Countries in economy, institutional system, technologies and mechanism has not been taken into proper consideration, therefore its interpretation has been restricted in many aspects. In fact, the economic and technological development in late-developing Large Countries is not balanced, demonstrating an obvious "diversified structure": some areas and sectors have the advantages of labor resources and suitable technologies, and some other areas and sectors have the advantages of capital and high and new technologies, and integrating these different advantages can form "comprehensive advantages of Large Country", it has the active role to promote regional economy coordinated development.[2]

The "diversified structure" in late-developing Large Countries is specifically demonstrated by the diversification of regions, of economy, of human capital and of technologies, and among them, the critical one is the diversification of technologies. In the development of regional economy, technologies are the fundamental driving force, and the root to produce the gradient in regional economy (Pione 1970). The "dual-economy theory" put forth by the economist Lewis is actually also a "dual-technology theory", he believed that the economic structure in developing countries can be divided into two sectors, that is, modern economy and traditional economy: the former is based on advanced sciences and technologies, with the characteristics of technology-intensive pattern; and the latter is based on traditional manual technology, with the characteristics of labor-intensive pattern (Chandra and Khan 1993).

In the view angle of development economics, the root cause of diversified structure of technologies is the difference in production efficiency resulted from the unbalanced productive force and economy development level between different economic sectors and different regions in a late-developing Large Country, that is, the technological gap between developed and backward regions (Boeke 1953), the existence of the technological gap results in different levels of technologies in regions, and these technologies at different levels just meet the needs in the economic development in the late-developing Large Country. In fact, to a late-developing Large Country, the advanced level of technologies is not a main decisive factor, and the key lies whether the existing technologies match with the actual conditions in regional economy or regional industries, and whether they can promote industrial upgrading to the maximum extent, and the coordinated development of regional economy. To a certain extent, even outdated technologies can still promote the economic development in late-developing regions and play a role of promoting coordinated development of regional economy as long as such technologies are adaptive to the technological foundation of the industries, and have a proper technological gap from developed countries or regions. To be more objective, a more adaptive technology can better promote the coordinated development of relevant regional economy, this adaptability is a capacity, and we refer it as the adaptive capacity of technology. On the contrary, if an industry with relatively backward technological foundation blindly imports advanced technology before a certain technology adaptive capacity has been gained, it still cannot realize technological catch-up or industrial upgrading, and it has been proved by many facts and researches (Krugman and Tsinddo 1991; Barro and Sala I-Martin 1997) that, the technological gap must be appropriate, and only with appropriate technological gap plus technology adaptive capacity suitable to it, can late-developing Large Countries realize technological catch-up by importing advanced technologies. Blomstrom et al. (1999) also came to the similar view in their research of technological gap and technological catch-up, only with the difference that they took technology adaptive capacity as a threshold for later-mover technological catch-up, but the essence of technological threshold is also the multiple adaptability of technologies. In China, Xu (2000) and Keller (2003) have also obtained similar conclusions, which support the view that technological diversity and adaptability are critical to coordinated development of regional economy.

All the above literatures state unanimously that, attention must be paid to technological gap in importing technologies in late-developing regions, the imported technologies must adaptive to the factor status in the regions to realize rapid coordinated development of regional economy. Basu and Weil (1998) further pointed out that, because of the differences between developing countries and developed countries in factor endowment, technologies in developed countries might not be surely suitable to developing countries, in this circumstance, if there is a big gap in factor endowment between the developing countries and developed countries, developing countries cannot make use of the technologies in developed countries to realize convergence to developed countries. The view of Basu and Weil is also applicable to different regions in Large Country economy, according to their understanding, so long as late-developing regions upgrade their factor endowment structure and narrow the gap from early-mover regions in factor endowment, they will be able to make better use of the developed technological to promote rapid economic growth, so as to realize economic convergence. However, they defined factor endowment only as per capita capital, which cannot effectively explain the reality in economic and social development. For example, many countries in Latin America, Africa and Asia have increased saving rate, but they have not achieved increased economic growth rate, to realize economic convergence. Acemoglu and Zilibotti (1999) extended factor endowment to the structure difference in human capital, i.e. the ratio of skillful labor to non-skillful labor. They pointed out that, as the cutting-edge technologies used in late-developing regions are invented in developed countries and regions, such technologies match with the factor endowment in developed countries and regions, but late-developing regions are lacking in skillful labor or human capital suitable to them, therefore, late-developing regions cannot effectively realize coordinated development of regional economy even if they import advanced technologies from developed regions.

The existing literatures only pay attention to the differences in factor endowment between regions, with a view that differences in factor endowment are critical to the effect of technology import in late-developing regions, and also to the coordinated development of regional economy, but they pay no attention to the effect of the extent of technology diversity on the import of technologies in late-developing regions in Large Countries, nor to the effect of the imported technologies on technology adaptability and adaptive capacity. The purpose of this chapter is to explore on the effect of technology diversity and adaptability on the regional economy coordination in late developing Large Countries by taking factor endowment and technology adaptive capacity as an endogenous variable, on the basis of the researches conducted by Romer (1990) and Barro and Sala I-Martin (1997). The framework of subsequent parts of this chapter is as follows: Part II discusses the relations between diversified adaptive technologies and the regional technology adaptive capacity in late-developing Large Countries, the technology adaptive capacity in late-developing Large Countries and regions is endogenous in the selection of suitable technologies in late-developing regions, instead of an exogenous variable. The technology adaptive capacity in late-developing regions will be high and economic convergence can be realized rapidly if the technologies imported in late-developing regions are adaptive to the existing technological foundation, human capital and material capital and no much efforts of learning is required. In Part III, the "leading—following" model is borrowed and the adaptive capacity factor is added to build a dual technological convergence model, to study the internal relations of technology adaptive capacity in promoting the coordinated economic development in late-developing regions of late-developing Large Countries. Part IV studies the adaptive capacity of diversified technologies with the human capital, material capital, technological foundation and industrial structure, and the functioning mechanism to promote the economic growth in late-developing regions. Part V is the conclusions and policy recommendations.

8.2 Regional Technology Adaptive Capacity in Late-Developing Large Countries

Technology diversity and technological adaptability are critical to coordinated development of regional economy, however, to a late-developing Large Country, diversification of technologies is after all not its ultimate goal, a late-developing Large Country must realize technological catch-up and technological convergence as quickly as possible with the technological adaptability resulted from diversification of technologies. This "technology diversified" characteristics and "technological convergence" task place its economic development in a unique reality: in a longitudinal view, it is in the transformation from a developing country to a developed country, some industries have completed transformation and realized technological upgrading, becoming the front of technological innovation in the country, for example the aerospace industry and laser industry in China, and some industries have not realized technological upgrading, still staying at a shallow level stage, even within the same industry, the technological level is different in different areas. This "technology diversified" period is one with traditional economy and modern economy, extensive economy and intensive economy coexisting; in a transversal view, between different areas, different sectors and different industries, the technological diversification and technology adaptive capacity are also not identical, and even within the same region, same sector and same industry, the technological levels are also in the trend of diversification. Given the reality of "technology diversity" and the task of "technological convergence", late-developing Large Countries must base themselves on "technology diversity" and "technological convergence" in formulating their economic development strategies, and full consideration should be given to the "integrated" advantages formed by "technology diversity", to formulate regional economic development strategies and policies starting from the realistic features of regional differences and technology diversification in late-developing Large Countries.

To late-developing Large Countries, especially those like China, India and Brazil, differences in regional factor endowment and economic development conditions have formed regions and industries with different technical conditions in the country, plus the differences in these different regions and industries due to factor endowment, human capital, material capital, industrial structure and regional economy development environment of different levels are formed. In the course of economic development, with the rise of human capital level, enrichment of material capital and optimization of industrial structure, the regional factor endowment can be gradually upgraded, becoming adaptive to the technical conditions of these regions and industries, and the more adaptive, they can produce more effect in the development of these industries, to promote the regional economy to develop at the highest speed, thereby gradually converging to regions with fairly high economic level within the late-developing Large Countries. In the meantime, in these late-developing Large Countries, due to the differences in human capital, material capital and regional economy development environment, some regions and industries have the conditions of early development, especially, guided by specific development strategies and policies, it can be ensured that some industries or regions can develop first to a certain extent. The development of such industries will guide the upgrading of human capital, enrich material capital and daily improve the economy development environment, to lay a foundation for the upgrading of other industries and the economic development in other regions.

In summary of the above, the technology adaptive capacity in late-developing Large Countries is not exogenous, instead, it is endogenous in the diversified technologies and factor endowment in late developing Large Countries. In the view angle of factor endowment, the differences in factor endowment within regions in late-developing Large Countries have decided the different levels in technologies between regions, resulting in differences in the regional economic development and conditions of industries. The factor endowments, especially the human capital, material capital and economic development environment, are just adaptive to the technical and economic conditions in different regions and different industries, or the diversified technologies can just give full play to the respective technology adaptability, enabling different industries to realize convergence as quickly as possible. Because for a specific region and industry, technologies either too advanced or too outdated are not good to the technological convergence in the region or industry, only technologies adapted to technology adaptive capacity can realize convergence with the technology spillover effect in developed regions and the imitative innovation effect in backward regions. If, with the intervention by government, technology adaptive capacity in backward regions can be fostered by policy guidance and stimulation, the gap in regional economy can be narrowed at accelerated pace, technological gaps can be eliminated more quickly, and technological convergence and economic convergence can be finally achieved. In other words, backward regions in late-developing Large Countries can import suitable technologies from developed regions, so that different factors can bring into maximum play their functions and effect, to narrow the technological gap, speed up technological innovation and promote convergence of late-developing regions to developed regions (Fig. 8.1).

发展中大国 Developing Large Countries
充足且较高水平的人力资本丰富的物质资本 Sufficient and high level human capital
Rich material capital
完善的经济发展环境 Complete economic development environment
匮乏且较低水平的人力资本 Scarce and relatively low level human capital
缺乏的物质资本 Insufficient material capital
不够完善的经济发展环境 Incomplete economic development environment
大国综合优势促进区域经济协调机制 Mechanism of CAOLC promoting coordination of regional economy
技术状况 Technical condition 1 Technical condition 2
区域经济、产业发展状况 Development conditions of regional economy and industries
区域经济、产业发展状况 Development conditions of regional economy and industries
多元性 Diversity 适应能力 Adaptive capacity 技术经济收敛 Technological and economic convergence
要素禀赋多元 Factor endowment diversity技术多元性 Technology diversity
技术适应性 Adaptability of technologies 区域或产业收敛 Regional or industrial convergence
技术溢出 Technology spillover 模仿创新 Imitative innovation
发达区域或产业 Developed region or industry 区域经济协调 Regional economy coordination
落后区域或产业 Backward region or industry

Fig. 8.1 Diversified adaptive technologies and coordination of regional economy in late-developing Large Countries

8.3 Multi-technology, Adapting Capacity and Regionally Coordinated Economic Development of Large Countries

To late-developing Large Countries, the existence of technological diversification leads to unbalanced development of regional economy. The fundamental of coordinated development of regional economy is the economic catch-up in late-developing regions, to realize a unitary structure of economy. The process of coordinated development of regional economy in late-developing Large Countries is one of late-developing regions catching up leading regions in late-developing Large Countries, only that this catch-up process can be more quickly realized with diversified technologies and adaptive capacity. Therefore, we add the adaptive capacity factor into the "leading–following" model of Barro et al. (1997), to analyze the process of coordinated development of regional economy in Large Countries. Suppose technology dualization exists in a late-developing Large Country, one is a region with relatively advanced technologies (A) and the other is one with relatively backward technologies (B), the technical potential difference resulted from the technical gap between regions can promote technologies to spread from region A to region B, and the late-developing region can bring into full play their later-mover advantage to fully absorb the spillover effect from the early-mover region. Suppose the technological innovation in region A is demonstrated by the quantity of intermediate input products N_A, the technologies developed by it (the intermediate input products) cannot flow freely, and will be mainly used to make end products in region A.[3] In region B, because of relatively low technological level, there is no matching human capital or material capital, so it is unable to invent intermediate products of similar technologies, and can only imitate or absorb products already invented and in use in region A. Suppose region B imports intermediate products, in a quantity of N_B, then N_A and N_B can represent the duality of technologies in the two regions. Because of fairly strong specificity of technologies, to adapt to the specific environment and products, region B needs to make appropriate transformation of the imported intermediate products, and this is the so-called imitation cost. According to the traditional theoretical hypothesis and realistic characteristics, it can be sure that imitation cost is obviously lower than innovation cost. Furthermore, because of the incomplete intellectual property right system in most developing countries, late-developing regions can usually obtain easily advanced technologies from early-mover regions, without paying high cost, that is, the weak intellectual property rights make region A unable to obtain any compensation from region B. With these preliminary hypotheses, let's define the production technologies and behavior of consumers.

1. Consumer preference

We suppose that the utility function of each household is:

$$U = \int_0^{\infty} \left(C^{1-\theta} - 1\right) / (1-\theta) \cdot e^{-pt} dt \tag{8.1}$$

Households make decisions and selection for utility maximization, where C is the per capita consumption, the population growth rate n is 0, θ is the marginal utility elasticity, and ρ the consumer preference. In assets, a household gets the rate of return *r*, and gets the wage rate *w* (equal to the marginal products of labor) on the total quantity of fixed labor L. The critical conditions required for household optimization is for consumption increase rate g, being

$$g = (r - \rho)/\theta \tag{8.2}$$

2. Analysis of market behavior of enterprises in developed regions

We convert technologies into intermediate input products to reveal the functioning mechanism of coordinated economic development in different regions. Following the ideas of Dixit and Stiglitz (1977), Romer et al. (1987, 1990), we write the production function of the end product made by a representative manufacturer in region A as

$$Y_A = A_A L_{A_j}^{1-\alpha} \cdot \sum_{j=1}^{N_A} \left(X_{A_j}\right)^a \tag{8.3}$$

where, Y_A is the output of end product, L_{Aj} the labor input or human capital input, j stands for the type of intermediate input product, $A_A > 0$ is the productivity parameter, representing the technological level of region A to a certain extent, N_A stands for the quantity of intermediate products in region A, and X_{Ai} for intermediate input products. Also $0 < a < 1$, and the additivity of $(X_{Aj})^a$ means that the marginal output of intermediate input products j is independent of the quantity of intermediate products used. Under normal conditions, the marginal output of intermediate input products is independent, which implies that the invention of a new product will not turn any existing product out of date. In the meantime, to study the effect of the quantity of intermediate products N_A[4], it is supposed that the intermediate input products can be measured with a common material unit, and the quantities used are the same, here it is supposed that for all intermediate input products $X_{Aj} = X_A$, so from formula (8.3), we obtain:

$$Y_A = A_A L_{A_j}^{1-\alpha} \cdot \sum_{j=1}^{N_A} \left(X_{A_j}\right)^a = A_A L_{A_j}^{1-\alpha} \cdot N_A \cdot X_A^{\alpha} = A_A L_{A_j}^{1-\alpha} \cdot (N_A X_A)^a \cdot N_A^{1-\alpha} \tag{8.4}$$

In fact, formula (8.4) means that intermediate input products can be measured with a common material unit, and the quantities used are equal, setting a common material unit for these intermediate input products is equivalent to that only one factor is input in the production process, only that N_A units are required for that factor. Formula (8.4) also shows that, for the given quantities of L_{Aj} and $N_A X_A$, the term Nl_A^a in the formula indicates that Y_A increases with the increase of N_A, which has grasped the essence of technological innovation, that is, technological innovation has avoided diminishing marginal return, reflecting the characteristic of endogenous growth of the production function.

We suppose that products from all enterprises are identical in kind, the output can be used in consumption, the production of intermediate input products X_A and the innovation input required in inventing new intermediate products (the additional N_A), and we measure all prices in the unit of the homogeneous product flow Y. So the profit of one end produce maker is:

$$\pi_A = Y_A - wL_{A_j} - \sum_{j=1}^{N} P_j X_{A_j} \tag{8.5}$$

where w is the wage rate, and P_j the prices of intermediate input products. As these producers are competitive, therefore they take w and price P_j as given, then formula (8.3) is differentiated

$$\frac{\partial Y_A}{\partial X_{A_j}} = A_{A\alpha} \cdot L_{A_j}^{1-\alpha} \cdot xL_{A_j}^{1-\alpha} \tag{8.6}$$

And (8.5) is differentiated with respect to intermediate input products X_{Aj} and labor L_{Aj}, and let it be equal to 0, so we obtain:

$$X_{Aj} = \left(A_{Aa}/P_j\right)^{1/(1-a)} \cdot L_{Aj} \tag{8.7}$$

$$w = (1—a) \cdot \left(Y_A/L_{Aj}\right) \tag{8.8}$$

To speed up the economic development, late-developing regions must energetically foster or import factors suitable to it such as material capital and human capital, to increase the innovation input of intermediate products. Suppose η units of Y should be input to innovate with a new product, that is, the cost of inventing a new product will not change continually. To stimulate innovation and speed up economic convergence, compensation should be made for successful technological innovation in a certain way. This is actually caused by the nature of public product of new technologies, and we also follow the traditional analysis thinking, to give innovators the patented property right system as incentives to technological innovation.

Suppose that after the invention of a new technology, the production cost of the jth intermediate product is Y per unit, therefore the present value of remuneration obtained from the jth intermediate product is:

$$V(t) = \int_0^\infty (P_j - 1) \cdot X_{A_j} \cdot e^{-\bar{r}\cdot(v,t)\cdot(v-t)} dv \tag{8.9}$$

where, $(\bar{r}) \cdot (v,t) \equiv [1/(v-t) \cdot \int_t^w r(w)dw]$ is the average interest rate from time t to u. If the interest rate is equal to a constant r (which can be proved), then the present value factor can be simplified as $e^{-\bar{r}\cdot(v-t)}$. The formula above indicates that, only when the selling price P_i exceeds the marginal production cost 1 for at least part of the time after period t, the fixed cost η for innovating with a new product can be compensated.

To a representative enterprise in region A, its market behavior of producing intermediate input products is restricted by $\pi_j = (P_j - 1) \cdot X_{Aj}$, and according to the requirement of profit maximization, it can be obtained:

It can be solved to obtain the price of intermediate product as:

$$P_j = P = 1/a > 1 \tag{8.10}$$

This shows that the price of intermediate input product is not changed. By substituting formula (8.10) into (8.7), we can determine the total quantity of each intermediate input product as

$$X_{Aj} = X = (A_{Aa}^2)^{1/(1-a)} \cdot L_{Aj} \tag{8.11}$$

Substituting formulas (8.10) and (8.11) into (8.9) and by rearranging it, we can obtain:

$$V(t) = (A_{A\alpha}^2)^{1/(1-\alpha)} \cdot L_{A_j} \cdot (1/\alpha - 1) \cdot \int_t^\infty e^{-\bar{r}\cdot(v,t)\cdot(v-t)} dv \tag{8.12}$$

As technologies are the motive force in the economic growth of a country, all countries and regions energetically support technological innovation for economic convergence, to enhance their own capability of self-reliance innovation. Therefore, it can be taken that technological innovation market is freely accessible, and all enterprises or individuals can pay the innovation cost η to ensure the smooth proceeding of technological innovation. In fact it means that, according to the features of innovation market, it can be approximately regarded as a completely competitive market. So on this market, it reaches balance when $V(t) = \eta$.[5] That is:

$$\eta = L_A \cdot A^{1/(1-\alpha)} \cdot L_{A_j} \cdot 1/(\alpha - 1) \cdot \int_t^\infty e^{-\bar{r}\cdot(v,t)\cdot(v-t)} dv \tag{8.13}$$

It can be seen from the above formula that all are constants except for the integral, therefore the formula can establish only when the integral term is also equal to a constant. This requires that the interest rate r(t) = r.[6] And with this, the integral is simplified into 1/r, so formula (8.13) can be converted into

$$r_A = (L_A/\eta) \cdot A_A^{1/(1-a)} \cdot a^{2/(1-a)} \cdot (1/a - 1) \quad (8.14)$$

The optimality condition for consumers means that the growth rate of consumption is given by $g_r = (r - \rho)/\theta$. Substitute formula (8.14) into it, the consumption growth rate is:

$$g_A = (1/\theta) \cdot [(L_A/\eta) \cdot A_A^{1/(1-a)} \cdot a^{2/(1-a)} \cdot (1/a - 1) - \rho] \quad (8.15)$$

The total output level of region A can be determined from formulas (8.2) and (8.11):

$$Y_A = A_A \cdot L_A^{1-a} \cdot X^a N_A = A_A a^{2a})^{1/(1-a)} \cdot L_A N_A \quad (8.16)$$

Therefore, for the given LA, YA grows at the same rate as NA.

The consumption level must be subjected to the budget restriction of economy:

$$C_A = Y_A - \eta g N_A - N_A X_A$$

where $\eta g N_A = \eta N$ is the resources input into technological innovation, and $N_A X_A$ the amount spent on intermediate products. Substitute it into the three formulas (8.11), (8.15) and (8.16), and we obtain after simplification

$$C = (N_A/\theta) \cdot \left\{(L_A/\eta) \cdot A_A^{1/(1-a)} \cdot a^{2/(1-a)} \cdot (1/a - 1)[\theta - a(1-\theta)] + \eta\rho\right\} \quad (8.17)$$

The above formula indicates that, for the given L_A, C_A and N_A grow at the same rate g_A as shown in formula (8.15). In summary of the above, when region A is in an equilibrium state, Y_A, N_A and C_A all grow at a fixed rate g_A.

3. Analysis of market behavior of enterprises in backward regions

Similarly, suppose the production function of the end product made by a representative manufacturer in region B is written as

$$Y_B = A_B L_{B_j}^{1-\alpha} \sum_{j=1}^{N_B} (X_{B_j})^{\alpha} \quad (8.18)$$

Y_B, L_{Bj} and A_B have the same meaning as those in a developed region, where N_B refers to the quantity of products invented in region A and used in region B, to facilitate our analysis, it is supposed here that no technological innovation is conducted in region B, so the quantity can be $N_A \geq N_B$. To facilitate our analysis without losing generality, it is also supposed that when a late-developing region imports technologies from an early-mover region, no expense will be paid, but it should bear the expenses of digestion and absorption. In the real economy, because it is within a country, and close ties exist on economic, social and cultural aspects, a late-developing region can easily learn technologies from early-mover regions by learning by doing, flow of talent personnel and trans-regional investment. Region B has imported intermediate products in a quantity N_B, but whether these intermediate products can be effectively absorbed to promote the economic growth in region B depends on the technology adaptive capacity of region B, or the adaptability of the human capital, technological level and material capital of region B to the imported technologies. By following the processing by Shiyuan and Lin (2006)[7] on the technological progress in late-developing regions, we suppose the productivity parameter of region B is $A_B = L_{Bj} \cdot F(N'_B/N_A) \cdot N'_B \cdot A_{B0}$, where L_{Bj} stands for the human capital in the imitated input technology, $\left(N'_B/N_A\right)$ the technology adaptive capacity of the late-developing region, N'_B the advanced level of technology imported by the late-developing region, and the greater (smaller) N'_B indicates the more advanced (backward) technology imported by the late-developing region. It should be emphasized particularly here that, $N_A \geq N'_B \geq N_A$ stands for the comparison of the advanced level of technologies imitated by late-developing region, rather than a comparison in quantity, the technologies imported should be at the same advanced level with the technologies in early-mover region, and their advanced level can at most reach the cutting edge of technology in the early-mover region N_A; A_{B0} is the existing productivity parameter of the late-developing region, representing the original technological level in the region. Also, $F\left(N'_B/N_A\right) > 0$, $\lim_{N'_B/N_A \to 1} F\left(N'_B/N_A\right) = 1$. To facilitate our analysis without losing generality, it is assumed that $F\left(N'_B/N_A\right) = e^{\lambda(1-N'B/NA)}$, where $\lambda < 1$. In this case, the productivity parameter of region B becomes $A_B = L_{Bj} \cdot e^{\lambda(1-N'B/NA)} \cdot N'_B \cdot A_{B0}$, which shows that the productivity parameter in the late-developing region B depends on the advanced level of the human capital, technological foundation and imported technology of the region, or the adaptive extent of the imported technology. The increase of the advanced level of the imported technology can on one hand promote the technological level of the late-developing region, and on the other hand can expand the technological gap with the late-developing region, reduce the technology adaptive capacity, thereby hindering the technological progress and economic convergence in the late-developing region. Therefore, there is an optimum level for the imported technology. For this, the target can be expressed as: max $e^{\lambda(1-N'B/NA)} \cdot N'_B$. Therefore, the optimum advanced level of the imitated technology for developing countries is, when $N_A < \lambda N_A$, there is a big gap between the technology in the late-developing region at the time with the cutting-edge technology of early-mover region, so at that time, the adaptive technological level imported by the late-developing region is N_A/λ, and it can achieve the most rapid technological progress. On the contrary, if the imported technology is at an excessively advanced level, because of the relatively low human capital and technological foundation in the late-developing region, the technological adaptive capacity in the late-developing region will be low, resulting in slow technological convergence in the late-developing region, therefore in such a case, the productivity parameter in the late-developing region is $A_B = L_B \cdot \frac{e^{\lambda-1}}{\lambda} \cdot N_A A_{B0}$. Based on the above-mentioned deducing process and relevant definitions, we make $m = \frac{e^{\lambda-1}}{\lambda} \cdot N_A$, which represents the advanced level of the technology imported by the late-developing region. Then $A_B = L_B j \cdot m \cdot A_{B0}$ indicates that, when the technological gap between the late-developing region and early-mover region is relatively big ($N_A < \lambda N_A$), the late-developing region can import adaptive technologies from early-mover region, to realize technological and economic convergence by "walking quickly with small steps"; when the technological gap between the late-developing region and early-mover region is not big, the late-developing region can imitate the cutting-edge technologies of the early-mover region, and at this time, $N_A \geq \lambda N_A$, $A_B = L_{Bj} \cdot {}^{\lambda(1-N'B/NA)} \cdot N_A \cdot A_{B0}$. To be more representative and typical, we only consider the first case, that is, there is a big gap between the late-developing region and early-mover region.

As mentioned above, to imitate the advanced technologies of region A, region B should make adaptive improvement to the intermediate products to be used, and such improvement will need some cost v, and here $0 < v < \eta$. According to the analysis thinking for region A, suppose each intermediate product in region B represents an addition of $1/a$ on the unit production cost, the quantity of each intermediate product in region B is:

$$X_{Bj} = \left(A_{Ba}^2\right)^{1/(1-a)} \cdot L_B \tag{8.19}$$

The total output of region B, similar to that of region A (as shown in formula 8.16), and per capita output are respectively:

$$Y_B = A_B \cdot L_B^{1-\alpha} \cdot X^\alpha N_B = (A_B \alpha^{2\alpha})^{1/(1-\alpha)} \cdot L_B N_B \tag{8.20}$$

$$y_B = Y_B / L_B = (A_B \alpha^{2\alpha})^{1/(1-\alpha)} \cdot N_B \tag{8.21}$$

It can be found from formulas (8.16), (8.20) and (8.21) that:

$$\frac{Y_B}{Y_A} = \left(\frac{A_B}{A_A}\right)^{1/(1-\alpha)} \cdot \left(\frac{L_B}{L_A}\right)\left(\frac{N_B}{N_A}\right), \frac{y_B}{y_A} = \left(\frac{A_B}{A_A}\right)^{1/(1-\alpha)} \cdot \left(\frac{N_B}{N_A}\right) \tag{8.22}$$

Substitute $A_B = L_B j \cdot m \cdot A_{B0} (N_A < \lambda NA)$ into the above formula, we obtain:

$$\frac{Y_B}{Y_A} = (L_{Bj} \cdot m)^{1/(1-\alpha)} \cdot \left(\frac{A_{B0}}{A_A}\right)^{1/1-\alpha} \left(\frac{L_B}{L_A}\right)\left(\frac{N_B}{N_A}\right) \tag{8.23}$$

$$\frac{y_B}{y_A} = (L_{Bj} \cdot m)^{1/(1-\alpha)} \cdot \left(\frac{A_{B0}}{A_A}\right)^{1/1-\alpha} \left(\frac{N_B}{N_A}\right) \tag{8.24}$$

It can be seen from the above two formulas that, the output level of the late-developing region B depends on the technological gap (N_B/N_A, A_{Bo}/A_A) and human capital gap (L_B/L_A) between the two regions and the human capital of the late-developing region and the advancement of imported technologies (m). These factors are also critical ones influencing the technological catch-up by late-developing regions and the coordinated development of regional economy. As far as per capita output is concerned, the economic gap of the two regions is more embodied in the difference in their technologies and technology adaptive capacity. On this basis, we obtain the first proposition of this chapter:

Proposition 8.1 The coordinated development of regional economy depends on the extent of technological diversification and human capital gap of the two regions as well as the human capital and the advanced level of the imported technologies in the late-developing region.

Region B obtains the same profit from selling intermediate products as region A, and region B also has the same free access conditions for imitative innovation as in region A (as shown by formula 8.13): the present value of profit is equal to the imitating cost v. The same as (8.14), the rate of return for region B is:

$$r_B = (L_B/v) \cdot A_B^{1/(1-\alpha)} \cdot \alpha^{2/(1-\alpha)} \cdot (1/\alpha - 1) \tag{8.25}$$

Substituting $A_B = L_Bj \cdot rn \cdot A_{B0}$ into a relevant formula, the consumption growth rate g_B in region B is:

$$g_B = (1/\theta) \cdot [(L_B/v) \cdot (L_B \cdot m \cdot A_{B0})^{1/(1-\alpha)} \cdot \alpha^{2/(1-\alpha)} \cdot (1/\alpha - 1) - \rho] \tag{8.26}$$

In the same way, we can obtain the economic equilibrium growth route of region B, for the given L_B, Y_B, C_B and N_B grow at the same rate g_B as shown in formula (8.26). This also indicates that the economic converging speed of late-developing regions depends on the human capital, technological foundation and advanced level of imported technologies in the late-developing regions.

To late-developing Large Countries, because of differences in regional economy, only a small number of areas can become developed first, and more areas are relatively backward in economy, where the population size is big and the cost of imitative innovation is relatively low. According to these conditions and in conjunction with the two formulas (8.15) and (8.26), we can come to the second important proposition of this chapter:

Proposition 8.2 The economic converging speed of late-developing regions depends on the human capital, technological foundation and advanced level of imported technologies in the late-developing regions. When $\frac{\nu}{\eta} < \left(\frac{L_B}{L_A}\right)\left(\frac{L_{Bj} \cdot m \cdot A_{B0}}{A_A}\right)^{\frac{1}{1-\alpha}}$, $g_B > g_A$, the economy in late-developing regions of late-developing Large Countries can grow more rapidly, with more powerful capacity of economic convergence.

Because $g_B > g_A$, the technological gap of late-developing regions will narrow gradually, and the advancement of the technologies imported by late-developing regions will become more and more adaptive to the human capital and technological foundation in the late-developing regions, so that the technology adaptive capacity in late-developing regions will gradually increase, with the direct outcome that the gap between N_A and N_B will also narrow gradually, with N_B increasing more rapidly than N_A, and finally the two will be equal, that is to say, after the backward region has learned all new technologies of the developed region, the two will grow at the same speed, and their gap will remain. In late-developing regions of late-developing Large Countries, the original productivity parameter A_{bo} in the local area is normally lower than the productivity parameter A_A in developed regions, because enterprises in early-mover regions are in possession of massive human capital, advanced management experience and excellent corporate culture, i.e. $A_A > A_{Bo}$, however, with the increase of adaptability of the imported technologies, plus the functioning of the potential of human capital, the productivity parameter of enterprises in late-developing regions can possibly become higher than that of the enterprises in early-mover regions, i.e. $A_A < A_{Bo}$, however, in whatever circumstances, late-developing regions can never reach the per capita output and wage rate level of the early-mover regions. Therefore we obtain Proposition 8.3.

Proposition 8.3 If $A_A > A_{bo}$, then: y_B and w_B can respectively converge to a level lower than y_A and w_A; if $A_A < A_B$, then y_B and w_B can respectively converge to a level higher than y_A and w_A, that is to say, when considered in the view angle of per capita output and wage rate, the late-developing region will finally catch up with the developed region, however, although both regions will grow at the same rate finally, they can never converge to the same per capita output and wage rate level.

Proposition 8.3 demonstrates that, although the two regions will grow at the same rate finally, to a late-developing Large Country, late-developing regions can never converge to the same per capita output and wage rate level of the developed region, and this situation has determined that the economic development strategies must be different in developed and backward regions, specifically including selection of industries, enterprise development strategies and policy guidance by the government, etc.

8.4 Mechanism of Coordinated Development of Regional Economy in Late-Developing Large Countries

We can conclude with the above three propositions that, late-developing regions in late-developing Large Countries can give play to the roles of human capital, material capital and technological foundation in late-developing regions by importing adaptive technologies, to realize rapid economic growth in late-developing regions, so as to achieve coordinated development of regional economy. In conjunction with the above analysis, we now make discuss in detail the mechanism of the technological diversity and technology adaptive capacity of late-developing Large Countries promoting coordinated development of regional economy.

1. Adaptation of diversified technologies and human capital can give great play to the utilization efficiency of human capital and technologies in late-developing regions

According to the view of new growth theory, the differences in the regional economic development level in late-developing Large Countries can be attributed to the differences in technological level between these regions, or the technological diversity between them. In an open economy, the technological gap from developed regions can be gradually narrowed through trade, investment, flow of professionals, technology spillover and imitative innovation, to promote the transformation of technologies from a multiple to a unitary pattern. However, in real economies, late-developing regions cannot narrow their differences from developed regions in technological level through technological transfer and imitation. In China, despite of the strategies to develop the west, rejuvenate the old industrial base of northeast and have the central China to rise, the gap between east and west regions kept on expanding, in the final analysis, the cause is that there is no human capital adaptive to the technological and economic development in these regions. Especially, in relatively backward regions of late-developing Large Countries, the low level of human capital leads to lacking in the capability to absorb technologies. The technological conditions of a region must be adaptive to its human capital, only technological level adaptive to human capital can fully play its role to drive economic growth, and to promote regional economy converging from a multiple to a unitary pattern. With a technological level adaptive to human capital level, a favorable dynamic cycle can be formed between human capital and technological level, thereby realizing sustained rapid development of economy.

2. Adaptation of diversified technologies and material capital can give great play to the utilization efficiency of material capital in all regions

Technological level cannot be brought into full play without human capital adaptive to it, however, human capital is only a carrier of technologies, and its full play must depend on certain material capital. According to definition, capital is a value that can bring about greater value than itself, and it is in two forms, the material capital and the human capital. The human capital, together with land comprised of natural resources and the material capital comprised of machines and equipment, are referred to as three factors in production, and their joint roles promote the rising of economic standard. To a certain extent, material capital and human capital can substitute each other, and human capital is often the embodiment of technological level. Especially, with the deepening and broadening of capital in the course of economic growth, there is a gradually enhancing tendency in the complementing or substituting relations between the material capital and human capital. More than that, investment also plays a promoting role to technological level and economic development through the material capital and human capital formed by it, and there is also a close tie between the material capital investment and technological level: the material capital investment is an important condition that technological level can be upgraded and can act upon economic growth. In the internal relations between material capital and technological level, technological progress is embodied in the material capital with machines and equipment as the main, or in the human capital with labor as the main, and "non-embodied technological progress" never exists; in technical production, the upgrading of technological level is also realized through material capital investment and human capital investment. In short, technological level must be adaptive to material capital, and act upon economic growth together with material capital. Therefore, only increasing technological level and human capital on one side without the matching with it by certain material capital, the high technological level cannot play its role, and vice versa. For this reason, diversified technologies must be coupled with diversified material capital to enable gradual convergence of regional technological level.

3. Adaptation of diversified technologies with regional economic development level can help upgrade the economic efficiency in regions

The coordinated development of regional economy in late-developing Large Countries is a process of optimizing and developing regional economy to a certain extent, and this requires that the diversified technologies are consistent with the overall level of development of regional economy. Technological level can rise gradually when technologies, with human capital as carrier, gradually absorb some knowledge, information and coding technique, and the transfer of knowledge and information requires certain economic development level, and certain environment for the development of regional economy. Only when the regional economy has developed to a certain level, can human capital start to flow, knowledge and information produce a spillover effect with the movement of the carrier, the regional economy level gradually change from multiple to unitary, and the gap in regional economy development be narrowed, to achieve economic convergence and coordinated development of regional economy. In the meanwhile, only when the technologies are adaptive to the regional economy development level, can the initiative of factor endowment in the region be motivated, to adapt factors of different quality to the technological level, and promote the efficiency of regional economy development to the maximum extent.

4. Adaptation of diversified technologies with industrial structure can speed up the industrial structure upgrading in all regions

In late-developing Large Countries, there are labor intensive industries, as well as capital-intensive industries and technology-intensive industries, and there are great differences in the development level and regional distribution of these industries. The economic practice in many developing countries has shown that, the most important barrier hindering the adjustment of economic structure is technological level, and neither the natural resources nor the material capital conditions. If the industrial structure required by economic development is not adaptive to technological level, material capital and human capital cannot be fully applied and advanced technologies cannot be effectively adopted, hindering the role of technologies in upgrading industrial structure. In the era of knowledge economy, the technological level determines the formation and change of industrial structure. The development of industrial economy towards knowledge economy is the evolution not only of economic form, but also of industrial structure upgrading and modernization, and this evolution requires that laborers have the necessary educational level, experience and skills, as well as the technological level matching with it. Theoretically, it is believed that industrial structure and its transformation should be in line with the industrial distribution of human capital inventory, that is, the technological level of corresponding proportions in industries should create corresponding portion of added values. However at present, the technological level conditions in China has hindered the evolution of industrial structure, on one hand, the human capital inventory in massive labor is low, so they can only be employed in traditional industries with low technological content; on the other hand, professionals for high and new technology industries with high technical contents are short, resulting in the contradiction of structural surplus supply and structural shortage of demand of labor in the country; the root cause is that the technological level and industrial structure are not adaptive to the demand of optimized configuration of industrial structure, and no sufficient coupling has been realized between industrial structure and technology diversification, as a result, the due matching capacity of the diversified technologies has not been brought into full play. Therefore late-developing Large Countries should adopt rational measures on the basis of their own technological conditions, to ensure matching of suitable technologies with suitable industries and the coupling of technological level with industrial structure, so that technological level of different conditions can adapt with different industries, in this way, it is possible to promote to the maximum extent the upgrading of industrial structure in late-developing regions and realize effective coordination of regional economy.

In brief, only when the diversified technologies are adapted to the human capital, material capital, economic development level and industrial structure in a region, can the role of adaptive technologies and the effect of factor endowment in the region be brought into full play, to promote the transformation from early-mover advantage to competitive advantage in developed regions and from later-mover advantage to comparative advantage in late-developing regions, so that late-developing regions can achieve autonomous and transcendent development, to realize coordinated development of regional economy. Also, the adaptation of diversified technologies to diversified factor endowment and regional development conditions can ensure the diversified technologies and regional factors foster competitive advantage while giving full play their comparative advantage, so as to realize coordinated development of regional economy.

8.5 Conclusions and Policy Recommendations

Late-developing Large Countries are characterized with a typical diversified structure for causes on many aspects, for example, the regional differences in production factors, marketization level, investment efficiency and urbanization level, however, different regions in late-developing Large Countries and even different industries in the same region have different production factors such as capital and technologies, and these result in gaps in development capacity and development level, the core of this gap is the diversification of technologies, and the diversity and adaptation of technologies are critical to coordinated development of regional economy, and also the source of comprehensive advantages of late-developing Large Countries. On this basis, we obtain the first conclusion of this chapter.

Conclusion 1 The technological adaptive capacity in late-developing regions of late-developing Large Countries is not an exogenous variable, instead, it is endogenous in the selection of suitable technologies in late-developing regions. The technology adaptive capacity in late-developing regions will be high and economic convergence can be realized rapidly through learning if the technologies imported in late-developing regions are adaptive to the existing technological foundation, human capital and material capital.

It is concluded in this chapter through analysis that the coordinated development of regional economy is determined by the technological gap and technology adaptive capacity of different regions, and to late-developing Large Countries, technological gap is a manifestation of technology diversification, and technology adaptive capacity is actually the capacity of technologies to adapt to relevant industries, regions and enterprises, therefore, in conjunction with Proposition 8.1, we come to the second conclusion of this chapter.

Conclusion 2 The coordinated development of regional economy in late-developing Large Countries is determined by the extent of technological diversification and human capital gap of the two regions and the advanced level of the human capital and the imported technologies in the late-developing region, and the speed of economic convergence in late-developing regions is determined by the human capital, technological foundation and advanced level of the imported technologies in late-developing regions.

Technologies are the motive power for economic development, however, it is not that the more developed technologies the better, and the key is that they should be suitable to the reality of local economy. Only technologies suitable to the technological foundation and technology adaptive capacity can really give play to their motive power mechanism, to promote the coordinated development of regional economy. It is a clear proof that many countries, regions and enterprises blindly import advanced technologies, but fail to realize technological convergence and catch-up. The root cause is that the imported technologies are not adaptive to the existing industries or existing foundation in the enterprises, and they were trapped into the vicious circle of "import–backward–import again–backward again". Therefore, in conjunction with Propositions 8.2 and 8.3, we come to the third conclusion of this chapter.

Conclusion 3 The fairly outstanding diversification in late-developing Large Countries can give full play to the adaptability of technologies, thereby the later-mover advantage of late-developing regions can be brought into play fairly well, to realize rapidly technological catch-up and coordinated development of regional economy. However, by only giving play to the technology diversity and adaptability, late-developing regions cannot converge to the same per capita output and wage level of developed regions.

Conclusion 3 indicates that, the coordinated development of regional economy in late-developing Large Countries are realized by various causes, the diversification and adaptability of technologies are critical to the coordinated development of regional economy, but it is not the sole decisive factor, and its realization also depends on many other factors, such as a healthy economic system and market system, and scientific and rational supporting policies. The policy guiding significance of this theory and conclusion in China is embodied in the following aspects:

First, China is a typical late-developing Large Country, so it can realize coordinated development of regional economy quickly as long as the effects of human capital, material capital and technological foundation in late-developing regions have been brought into full play. With the daily tightening ties in regional economy and increasing free flow of factors, full utilization of the spillover effect of advanced technologies from developed regions has become an important source of technological progress in late-developing regions, however, in importing technologies from developed regions, the advanced level should not be taken as the only criterion, we should base it on actual conditions, give full play to the adaptability of diversified technologies, create conditions to improve technological foundation, foster human capital, upgrade technology adaptive capacity and give full play to the adaptive capacity of diversified technologies to the maximum extent. Therefore, in formulating regional economy policies or industry and science and technology policies of the state, efforts should be made to foster the technology adaptive capacity of regions and the country, upgrade the technological foundation, learning ability and social capacity, increase the R&D input and education input, improve the education system, upgrade the R&D capacity, make full use of and absorb the technology spillover effect from developed countries or regions, fully tap the own foundation and quality and increase the adaptability of technologies in all aspects.

Second, China is a country with significant multiple characteristics, its economic development and technological level are quite unbalanced in the east, central and western regions, between the east coastal region and the central and western regions, there are not only great differences in learning ability and social capacity in labor, capital and technologies, but also in the development and completion of market institutional organization and systems. These conditions have determined high technology diversity in the regions, therefore, in importing technologies, regions should base on their actual conditions, including the human capital and material capital conditions and economic and social development environment, to import adaptive technologies, to allow the potential of technologies in different regions to be brought into full play, and realize rapid economic growth in late-developing regions. For example, in the east region, the human capital is fairly rich, material capital is relatively abundant and the technological foundation factor endowment has been greatly upgraded after years of development, therefore, this region must import technologies at fairly advanced level, to ensure matching of the imported technologies with the technology adaptive capacity in the region, so that the imported technologies can be fully digested and absorbed and secondary innovation can be launched, to enrich the connotation of economic development in the east region and realize the transformation from factor driven growth to technology driven growth. The central region, after receiving technology transfer for years, have built up a fair level of factor endowment, so it should import suitable technologies of medium level, to ensure that the imported technologies can be fully digested and absorbed and combine with the factor endowment in the region, so as to bring into full play the functions and effect of factors and promote the rapid development of economy in the central region. Presently in the west region, the material capital, human capital and technological foundation are relatively weak, therefore, suitable technologies must be imported in conjunction with its own actual factor endowment, especially, technologies adaptive to its rich natural resources should be imported, to give play to the effect of regional factors, promote the rapid development of economy in the west region, and finally realize the coordinated development of all east, central and western regions.

Notes

1. The late-developing Large Countries here refer to countries like China, India and Brazil with large national territorial area, rich resources, large population, and diversified development levels. These countries have the reality of technology diversity as well as regions and industries with different adaptability resulted from the diversified technologies, therefore these late-developing Large Countries have comprehensive advantages.
2. Yao Ouyang: *The Direct Outward Investment Strategy of China Based on Comprehensive Advantages of Large Country, Economics of Finance and Trade* 2006 Vol. 5.

3. This hypothesis is somewhat far-fetched, but it complies with the reality fairly well, because today with highly developed economy and technology, many intermediate input products in enterprises are tailor-made in some factories, especially, for some complicated technologies, parts are often supplied by many associated factories, and such tailor-made intermediate products or parts are highly dedicated to certain assets, and can be hardly used in making other similar products.
4. For convenience, the quantity of intermediate products can be understood as continuous, instead of discrete. In this way, N_A can represent the sophisticated level of technologies in late-developing regions or enterprises, and can actually prove the continuity nature of N in a formalized way, we can integrate N_A in formula (8.3): (公式), where j is the continuous indicator of quantity, and N_A the range of types of intermediate products that can be used.
5. Because, if $V(t) > \eta$, infinite factors will be input into the research and development market at the time t, and if $V(t) < \eta$, no corresponding factor will be input into innovation at the time t, therefore the number of products N will change with time. Therefore, at equilibrium state, $V(t) = \eta$.
6. This result can also be obtained this way: differentiate the integral (公式) with respect to t, and the result is $dI/dt = -1 + r(t) \cdot I - 0$. Therefore $r(t) = r$, $I = 1/r$.
7. Here, the idea of Shiyuan (2006) was borrowed in processing the productivity parameter of backward region, but with the difference that we take into account the adaptability of the human capital, material capital and technological foundation in late-developing regions with the imported technology, instead of the technological progress.

Chapter 9
Case: Outward Investment Strategy of Late-Developing Large Countries

This chapter analyzes the outward investment strategy based on comprehensive advantages of Large Country with outward investment as a case. China is a late-developing Large Country, which should be the basis for analyzing economic development strategy as well as analyzing outward direct investment strategy. The comprehensive advantages of Large Country reflect our basic national conditions, and determine the characteristics of our outward direct investment in industries, regions and the selection of subjects. We should say that the outward direct investment strategy based on comprehensive advantages of Large Country is a scientific strategy on foreign economic relations and trade that complies with the national conditions. The outward direct investment of China is "resources integrated", it should integrate the natural, economic and technological resources of different regions and at different levels, to form the comprehensive advantages of Large Country combining the advantages of both developed countries and developing countries as one, to obtain best interests from outward direct investment.

9.1 Related Literatures

At the beginning of the 21st century, with the proposing and implementation of the "going global" strategy of China, outward investment became a hot spot of research in the theoretical circle. According to many scholars, China, as a late-developing Large Country, implements outward investment for the purpose of seeking for resources development, driving up export trade, opening overseas market, evading trade barriers, obtaining high and new technologies and establishing outward windows, and in essence, it is a process to make use of, maintain, give play to, develop and seek for advantages. Sun (2000) summarized the comprehensive

Y. Ouyang, *The Development of BRIC and the Large Country Advantage*,
DOI 10.1007/978-981-10-0633-3_9

advantages of China's outward investment as multi-polarization of investment motivations, diversification of differential advantages and development spaces, and analyzed their internal logic relations, the mutual stimulation between multiple development targets, the multiplication effect between multiple factors of differential advantages and actuation mechanism between the co-existing multi-stages. Lu (2003) analyzed the advantages of Chinese enterprises in transnational operations, that is, the later-mover advantage from "imitative innovation" activities, the cost advantage from inexpensive labor, and adaptive technology advantage from the practical use and localization of technologies. Song (2001) summarized five advantages with the outward investment of China: the comprehensive scale advantage adapting to different investment demands, the advantage of mature technologies and traditional products, the competitive advantage of low product costs, the talent advantage suitable to internationalized operations, the racial link advantage and good domestic and international environment.

Some scholars analyzed the advantages of different types of enterprises in outward investment, mainly the advantages of large and medium-sized enterprises, small enterprises and privately run enterprises. Wang (2004) stated that large and medium-sized transnational companies have the advantages of ownership and internalization, and small enterprises have the advantages of small scale technology and localization of technologies. Zhou (2004) analyzed the ownership advantage of privately run enterprises in outward investment, including the technology advantage in developing labor intensified and technology intensified composite products, the marketing advantage of direct transaction, the low-cost management advantage, and the unique advantages of flexible and autonomous operations and high degree of marketization. Ouyang (2005) stated that the characteristic of privately run enterprises with clearly defined property rights has determined their advantage of operation autonomy and flexible mechanism, which lead to efficiency advantage and relevant technological advantages, and they constitute as a whole the unique advantage of privately run enterprises in outward investment. Fan (2004) analyzed the role of clustering and strategic technological unions of small and medium-sized enterprises in upgrading the transnational competitiveness, believing that the advantages of enterprise groups are embodied on production cost, regional marketing, technological innovation, core capabilities, production efficiency and incentive competition, and the effect of strategic technological unions of enterprises are embodied in favoring complementation of resource advantages and upgrading core technical competence, conducive to saving the learning and cooperation cost and increasing the learning efficiency in enterprises, conducive to the establishment, obtaining and development of relationship resources among enterprises, and conducive to speeding up the access of technologies into market.

About the selection of outward investment strategies, domestic scholars have put forth valuable ideas and conceptions. Jianguo (2003) put forth the hypothesis of effective capital based international direct investment, that is, under the condition of economic globalization and free market at dominating position, the FDI capacity of enterprises in developing countries is fostered gradually in the course of internationalization of the value-adding chain of the industry, and this process is in essence fostering the adaptability of enterprises to transnational operation environment, that is, enterprises, through adaptive integration of production factors, transform effective domestic capital into effective international and global capital. Wang (2004) stated that China needs to participate in more development of overseas natural resources, to make use of overseas scientific and technological resources, to drive up export through outward investment, to get closer to overseas market, and to promote adjustment of domestic industrial structure. On selecting the subjects of outward investment, Xu (2002) advocated "encouraging large enterprises while relaxing control over small ones", that is, major enterprises, large and medium-sized enterprise groups and listed companies with powerful competence, good management and brands of their own should be encouraged to expand their overseas investment, and small and medium-sized enterprises and privately run enterprises with good economic performance and flexible operation mechanism should also be encouraged to run factories overseas. In the selection of industries, Cheng (2001) thought that arrangement for investment to obtain resources should be taken into consideration first, followed by the advantages of traditional industries, and then the tracking of high-tech industries in developed countries, and Guoyuan (2008) stated that attention should be paid to resources exploiting industry, labor-intensive industries and those applying mature and suitable technologies should be developed, and investment for seeking technologies should be increased. Chen and Feiqiong (2008) proposed that, in location choice for outward investment, countries with economic development little bit slower that China should be selected; and joint venture should be the main form of outward investment, as this can better obtain advanced technologies of other countries and use foreign investment. On the basis of analyzing the result of outward direct investment by China, Li (2009) advocated expanding the scale of investment in underdeveloped countries, to speed up the gradient industrial transfer; in developed countries, investment in technology-intensive industries should be increased concurrent with appropriate transfer of gradient industries; selection of regions and industries should be optimized, investment in Africa and the Middle East should be promoted actively, and the investment proportion in manufacturing industry be increased.

In summary of the above, domestic scholars have studied the outward direct investment strategy of China from different view angles, and put forth corresponding countermeasures and recommendations based on the realistic situations of development of international economy and Chinese economy. However, there are two defects in the current researches: first, systematic research is lacking on the advantages of outward investment by China; and second, no complete outward investment strategy has been proposed. In this chapter, the author attempts to propose a fairly complete strategic conception by studying the outward investment strategy of China on the basis of the principle of "comprehensive advantages of Large Country".

9.2 Outward Investment Strategies of China

The unbalanced economic development in Large Countries, with differences in resources and technological conditions, is conducive to integrating resources to realize complementation of advantages. Outward direct investment should be backed by the advantages in parent country, China, as a late-developing Large Country, has advantages with obvious "differences" when viewed partially, and also has "comprehensive" advantages as a whole. Such advantages of "differences" and "comprehensiveness" can be used as a backup, to implement "differential" outward direct investment by selecting different industries, regions and enterprises, to build a "comprehensive" outward direct investment strategic framework.

1. Selection of industries based on comprehensive advantages of Large Countries

The selection of industries for outward direct investment by China should be based on the comprehensive advantages of Large Country, that is, be established on the characteristics of regional, economic, technological multi-structures. First, China is a Large Country, and should establish a complete industrial structure in coordinated development, therefore its outward direct investment should also select industries rationally, to facilitate the adjustment and upgrading of industrial structure. Next, China is a developing country, and its economic development and industrial development level is not high, therefore in outward direct investment, it should give full play to the advantages of small-scale technologies and suitable technologies, and select some traditional industries, featured industries and labor intensive industries with advantages. Third, the economic and technological development in China is unbalanced, some high and new technology industries have been well developed, therefore, in outward investment, these advantageous industries should not be neglected, and high and new technology industries should be made as the main direction in development and be energetically fostered and supported.

The Chinese Government has stated that "we will promote optimization and upgrading of industrial structure, and form an industrial pattern led by high and new technology industries, supported by basic industries and manufacturing industry with service industry developing in an all-round way", and it is also the industrial pattern that should be formed in outward direct investment; in the meanwhile, it is stated that "we should correctly handle the relationship in developing the high and new technology industries and traditional industries, capital and technology-intensive industries and labor intensive industries, virtual economy and physical economy", and this is also an important principle in selecting industries in the outward direct investment of China. Specifically, there are mainly two aspects: first, we should correctly handle the relationship in developing high and new technology industries and traditional industries. The conditions in the late-developing Large Country have determined the dual tasks in the industrialization of China—we should accomplish not only the task of traditional industrialization, but also the task of catching up with the new industrial revolution in the world. China is currently still in the medium phase of industrialization, and traditional industries still have broad prospects of development. In the meantime, the high and new technology industries, as leading industries in the process of industrialization, are also developing rapidly in China. Therefore in outward direct investment, we should base on traditional industries and also give priority to the development of high and new technology industries. In underdeveloped regions, forces should be concentrated on developing outward direct investment with traditional industries; in developed regions, emphasis should be placed on developing outward direct investment with high and new technology industries. Today, scientific and technological enterprises "going global" has become a new trend, for example, the Haier Group, Huawei Group and Goldtel Information Group all went to the international market with their advantages of high and new technologies. Second, we should correctly handle the relationship between developing capital and technology-intensive industries and developing labor intensive industries. China has a hierarchic economic structure, and labor intensive industries still have great potential of development. We have rich human resources and fairly low labor cost, which is a unique advantage in international economic competition and must not be neglected in outward direct investment. In the meantime, developed regions and high grade enterprises should energetically develop outward direct investment in capital and technology-intensive industries.

2. Selection of regions based on comprehensive advantages of Large Countries

China is a Large Country with unbalanced internal economic development, with different economic and technological levels in different regions and enterprises, and the resources conditions and market demand are not identical in various places. Therefore, in selecting areas for outward direct investment, we should also base ourselves on the characteristics of comprehensive advantages of Large Country, and analyze and treat specific issues on a case by case basis. First, big enterprises can choose to invest in developed countries, and small and medium-sized enterprises can choose to invest in developing countries. Some big enterprises or enterprise groups have advantages of abundant finance, advanced technologies, management talents and known brand products, with the conditions to compete with enterprises in developed countries, therefore, by investing in developed countries, they can make use of not only the sound economic environment in these countries to foster and mould their own operation mechanism, but also the R&D centers in developed countries to seek for international cutting-edge sciences and technologies, to enhance their international competitiveness. Some small and medium-sized enterprises have the advantages of clearly defined property rights, flexible mechanism, adaptive technologies and fairly low cost, but they don't have the scale and financial competence to invest in developed countries, so they can invest in developing countries, to make marketable products with the small scale production technologies and production experience and the natural resources in developing countries, and open up market in developing countries, for better economic gains. Second, high and new technology industries can choose to invest in developed countries, and traditional industries can choose to invest in surrounding countries.

Developed countries have the superior conditions and ideal environment to develop high and new technologies, suitable to investment by high and new technology industries. High and new technology industries making investment in developed countries is conducive to their research and development, their catch-up with international cutting-edge sciences and technologies, and their own growth and development. Traditional industries should mainly invest in surrounding countries, because the consumers in these countries have the traditional customs and consuming habits similar to those of China, therefore there are similar market demands, giving a special market environment advantage to the traditional industries of China.

Of course, some special cases should also be taken into consideration. For example, big enterprises and high and new technology industries of China can also make investment in developing countries, to gain advantages on scale, technology, marketing and monopoly, and obtain high monopoly profits, also, to avoid opposition from developing countries, joint venture operation can be implemented with enterprises in developing countries. On the other hand, small and medium-sized enterprises and traditional industries of China can also invest in developed countries, as small and medium-sized enterprises can build up scale advantage by "clustering" to enter the market of developed countries, or seek for development space on some links in the international industrial chain, to obtain their value from international division of labor; traditional industries can enter the Chinese communities in developed countries, to meet their demands for traditional products, and can also create demand among consumers in developed countries through publicity and promotion, to open up markets in developed countries.

3. Selection of enterprises based on comprehensive advantages of Large Countries

Enterprises in China differ greatly in scale and capital competence, as well as in technology and management level, also different types of enterprises have different advantages, which are suitable to different industries, different areas and different investment directions. Because of this characteristic of comprehensive advantages of Large Country, we should make scientific and rational selection of microscopic subjects.

On one hand, we should foster large enterprises and organize transnational enterprise groups, to give play to the advantages of group operations. The important position of enterprise groups in outward direct investment is determined by their overall quality and advantages, such as their advantages on internalization and resources integration. On the international market with monopolized competition, we face big enterprise groups, and even "unions" of large transnational groups as competitors, to contend with these transnational big enterprises and consortiums, we must bring into play the role of our large enterprise groups, and select some enterprise groups to give them major support, to form industry and trade integrated large transnational groups, and they should endeavor to rank among the global top 500, to become the main bodies for transnational operations by Chinese enterprises.

On the other hand, we should encourage and support small and medium-sized enterprises to "go global", to form enterprise groups with industries as links, and demonstrate their advantages of group operations. Enterprise clustering is formed by companies and entities with mutual links in specific areas and theoretically close to each other, they have the advantages in production cost, regional marketing and technological innovation, therefore on a favorable position to obtain new and complementary technologies, speed up the process of learning, reduce transaction cost, form marketing networks, overcome or build market barriers, obtain synergic economic benefit and divert risks. It is inevitable to a certain extent that small and medium-sized enterprises form clusters in transnational operations, as this can maintain the advantages of small and medium-sized enterprises in flexible mechanism and adaptation to small scale market, and also gain scale advantage, therefore they can concurrently have some advantages of small and medium-sized enterprises and big enterprises. The small and medium-sized enterprises in Zhejiang have formed industrial clusters with distinctive features in the form of "agglomeration", thereby increasing their integrated competitiveness to participate in international competition, such as the clusters formed by garment and metal ware producers, demonstrating obvious advantages in transnational operations. Especially, the small and medium-sized enterprise clusters formed for the direct procurement by transnational companies have unique advantages on absorbing technology spillover: many enterprise groups making automobile parts in Hangzhou, Taizhou and Wenzhou have entered the division of labor system in the industries of developed countries by making associated automobile parts for transnational companies such as Ford, Delphi and Iveco, gradually mastering the advanced production technologies and taking some of the international OEM market and after-sale service market of automobile parts.

9.3 International Comparison of Outward Investment Strategies

The present situation shows that the "comprehensive advantages of Large Country" in connection with the diversified regional structure, economic structure and technological delivery is specific with developing Large Countries. They are different from developed countries and other developing small countries, but have integrated the features and advantages of developed countries and developing countries. Here, we will make comparison on the outward direct investment strategies of China with developed countries and other developing small countries.

1. Comparison of China with developed countries

Generally speaking, the outward direct investment of developed countries has the following main characteristics: first, outward direct investment of developed countries has the monopoly advantages on capital, technologies and scale, high input and high technologies produce high efficiency, plus the internalized transaction between parent companies and subsidiaries, large amount of transaction expenses can be saved. Second, the outward direct investment of developed countries is usually aimed at obtaining high profits, and the proportion of profit re-investment is also high. Third, the investment in developing countries is mainly aimed at utilizing the natural resources, labor resources and broad markets in developing countries. In the strategies of transnational companies of developed countries towards China, they attach most importance to two factors: one is the market attraction of China. The population of 1.2 billion means a huge market, and with the daily prospering economy, the purchasing power is increasing. The second is the labor force resources of China. In developed countries, the labor cost is high, while developing countries usually have large amount of inexpensive labor. Fourth, as the industrial structure for outward direct investment in developed countries is in the trend of weakening, they placed emphasis on high and new technology industries, knowledge-oriented service industry and traditional industries after transformation with high and new technologies.

Compared with developed countries, China has differences but also some identical aspects. On different aspects: first, the enterprises in China are in relatively small scale, and the advantages on economies of scale and internalization are not apparent, and second, the outward direct investment by Chinese enterprises is mainly to seek for technologies or market, without the motivation of seeking for labor resources. On the identical aspects, China is also taking the road of informatization driving industrialization, and the industrial structure of outward direct investment is also upgrading step by step. China is implementing the strategy of developing the IT industry and high and new technology industries in priority and transforming and upgrading traditional industries with advanced science and technologies. Since the mid 1990s, the high and new technology industries of China has been developing continually at a high speed, reaching advanced world level on some aspects, and they have become emerging pillar industries to drive export and economic growth, and also established competitive advantage of Chinese enterprises in outward investment and transnational operation.

2. Comparison of China with developing small countries

Generally, the outward direct investment of developing countries has the following features: first, mature and suitable technologies are adopted to develop traditional industries, suitable to small scale market. The suitable technologies of developing countries have the unique advantages not possessed by enterprises of developed countries in the outward direct investment to other developing countries. Second, featured products are developed with traditions and customs as links, to adapt to special market environments. Some featured products of developing countries can meet the consuming demands in surrounding countries with common traditions and customs, and these are things that enterprises in developed countries are not willing to do or cannot do. Third, the advantage of relatively low cost is used to save expenses on labor, management and marketing, to gain the advantage of lower prices. When investing in other developing countries or developed countries, developing countries mainly use the labor and technical personnel and management personnel of their own countries, so the prices can be relatively low. We can say that these features are general with developing countries, and the outward direct investment of China also has these features. For example, enterprises of outward direct investment from China produce textiles, garment, shoes, bicycles and some small electronic products, they are all in small scale, most of them based on labor intensive production techniques and processes; the outward direct investment in sectors of traditional Chinese medicine, fireworks and foodstuff has distinctive traditional ethnic features, mainly to meet the demands in overseas Chinese communities; the knitting company set up by Shanghai in Mauritius got the local market with the advantage of low cost, as the cost price of its products is only about one third that on European market, and it has also accessed to the EU market.

The outward direct investment of developing countries generally have the above-mentioned features, however, there are some differences between developing small countries and late-developing Large Countries: the former only have these features, while the latter also have some preliminary features of developed countries in addition to the above. China, as a late-developing Large Country, also demonstrates some features similar to those of developed countries in its outward direct investment: first, the outward direct investment of China is oriented to other developing countries as well as developed countries. Especially, with the upgrading of industrial structure and the growing motivation to seek for technologies, the growth of investment in developed countries will exceed that in developing countries. Second, the outward direct investment of China attaches importance to the development of traditional industries as well as the development of high and new technology industries. With the upgrading of development targets of enterprises, the transnational enterprises of China pay more attention to the research and development of technologies, for example, Haier and Huawei have increased their input to technologies and project research and development, and third, China has also started enterprise group operations, to obtain some advantages of internalization. The integrated operations of production and trade in transnational groups can save operation cost through the form of administrative organization and internal transaction within organization, embodying the internalization advantage of organization.

9.4 Analysis of Data of China and Policy Recommendations

According to the statistical bulletin on outward direct investment of China in 2009, by the end of 2009, 12,000 domestic investors of China had set up 13,000 outward direct investment enterprises in 177 countries and regions over the world, the

Table 9.1 Statistical results by years since China established the statistical system for outward direct investment (*Unit* 100m USD)

Year	Flow	Inventory	Year	Flow	Inventory
2002	27.0	299.0	2006	211.6	906.3
2003	28.5	322.0	2007	265.1	1179.1
2004	55.0	448.0	2008	559.1	1839.7
2005	122.6	572.0	2009	565.3	2457.5

Source: *2009 Statistical Bulletin of China's Outward Foreign Direct Investment*, the data of 2002–2005 are the statistical data of China's non-financial outward direct investment, and those of 2006–2009 are the outward direct investment data of all industries

outward direct investment inventory exceeded 245.75 billion US dollars and the total amount of assets in overseas enterprises exceeded 1 trillion US dollars.

The Table 9.1 show that the outward investment of China has been in steady growth, and the growth has accelerated in recent two years. Some scholars made empirical research using data at enterprise level on the output growth effect and technological upgrading effect in the outward investment of China, and the results show that in general, there is no difference in output growth and technological upgrading effect between overseas investment and domestic investment of enterprises. In a late-developing Large Country, such a result should be a fairly successful one, basically realizing the goal of obtaining overseas resources. In a more detailed analysis, the outward direct investment of China has the following features:

(1) In general, the outflow increases rapidly and the inventory scale is large. In 2008, the outward direct investment flow of China was 55.91 billion US dollars, a year-on-year increase of 111 %; in 2009, the global economic development came to a valley with the impact by financial crisis, and it also provided an appropriate time for Chinese enterprises to participate in international investment cooperation. In 2009, the outward direct investment flow of China reached 56.53 billion US dollars, hitting a historical high again. The growth of outward direct investment flow started to accelerate in 2004, and the annual average growth rate of outward direct investment from 2002 to 2009 reached 54.4 % (refer to Fig. 9.1). In the meanwhile, the inventory scale of outward direct investment kept on expanding, the growth accelerated starting from 2006, and it reached 245.75 billion US dollars in 2009 (refer to Fig. 9.2).

(2) In terms of industries, it covered all industrial sectors, with some outstanding major ones. China's outward direct investment covers many industries in a fairly complete range. In 2009, the distribution of China's outward direct investment flow by industries is: 20.47 billion USD in leasing, commerce and service industry,

对外投资流量 Outward investment (100m USD)

Fig. 9.1 China's outward direct investment flow during 1999–2009. Source: *2009 Statistical Bulletin of China's Outward Foreign Direct Investment*

对外投资存量 Inventory of outward investment (100m USD)
2002 2004 2006 2008

Fig. 9.2 China's outward direct investment inventory during 2002–2009. Source: *2009 Statistical Bulletin of China's Outward Foreign Direct Investment*

accounting for 36.2 %; 8.73 billion USD in banking industry, for 15.5 %; 6.14 billion USD in whole-sale and retail sectors, for 10.8 %; 13.34 billion USD in mining, for 23.6 %; 2.07 billion USD in communications, transportation, warehousing and post industry, for 3.7 %; 940 million USD in real estate, for 1.6 %; 780 million USD in scientific research, technical services and geological survey, for 1.4 %; 2.24 billion USD in manufacturing industry, for 4 %; and 470 million USD in power, coal and water production and supply industries, for 0.8 % (refer to Fig. 9.3).

In 2009, the distribution of China's outward direct investment flow by industries was as follows: 72.95 billion USD in business service industry, accounting for 29.7 %; 45.99 billion USD in banking industry, for 18.7 %; 35.7 billion USD in whole-sale and retail, for 14.5 %; 40.58 billion USD in mining industry, for 16.5 %; 16.63 billion USD in communications, transportation, warehousing and post industry, for 6.8 %; 13.59 billion USD in manufacturing industry, for 5.5 %; and 5.34 billion USD in real estate industry, for 2.2 % (refer to Fig. 9.4).

(3) In terms of regions, the investment has a wide coverage, with some outstanding regions. In 2009, China's outward direct investment mainly flowed to: China Hong Kong at 35.6 billion USD, accounting for 63 %; Cayman Islands at 5.37 billion USD, for 9.5 %; Australia at 2.44 billion USD, for 4.3 %; Luxembourg at 2.27 billion USD, for 4 %; British Virgin Islands at 1.61 billion USD, for 2.9 %; Singapore at 1.41 billion USD, for 2.5 %; and the United States at 910 million USD, for 1.6 %. In the distribution by regions, in 2009, the investment in Europe, North America and Latin America increased by times over the previous year.

Commerce and service industry
Mining industry
Banking industry
Whole-sale and retail
Manufacturing industry
Communication and transportation/warehousing and post industries
Real estate industry
Scientific research/ technical services and geological survey
Power/ coal gas and water production and supply industry
Building industry
Agriculture, forestry, animal husbandry and fishery
Information transmission/ computer service and software
Resident service and other services
Other industries

Fig. 9.3 Distribution of China's outward direct investment flow by industries in 2009. Source: *2009 Statistical Bulletin of China's Outward Foreign Direct Investment*. (*Unit* 100m USD)

Fig. 9.4 Distribution of China's outward direct investment inventory by industries in 2009. Source: *2009 Statistical Bulletin of China's Outward Foreign Direct Investment*

(100m USD)
Commerce and service industry
Banking industry
Mining industry
Whole-sale and retail
Communication and transportation/warehousing and post industries
Manufacturing industry
Real estate industry
Building industry
Technical services and geological survey
Power/ coal gas and water production and supply industry
Agriculture, forestry, animal husbandry and fishery
Information transmission/ computer service and software
Water conservancy/environment and public facilities management
Resident service and other services
Accommodation and catering industry
Other industries

The investment in Europe was 3.353 billion US dollars, increasing by 2.8 times compared with the previous year, accounting for 5.9 % of the total flow, and it mainly flowed to Luxembourg, Russia, the United Kingdom, Germany and Holland. The investment in North America was 1.522 billion US dollars, increasing by 3.2 times compared with the previous year, accounting for 2.7 % of the total flow, and it mainly flowed to the United States and Canada. The investment in Latin America was 7.33 billion US dollars, doubling that of the previous year, accounting for 13 % of the total flow, and it mainly flowed to Cayman Islands, British Virgin Islands, Brazil, Venezuela and Peru (refer to Table 9.2).

In 2009, the regional distribution of China's OFDI was as follows: 185.54 billion USD in Asia, accounting for 75.5 %; 30.6 billion USD in Latin America, for 12.5 %; 9.33 billion USD in Africa, for 3.8 %; 8.68 billion USD in Europe, for 3.5 %; and 6.42 billion USD in Oceania, for 2.6 %. There were 18 countries and regions each with investment inventory over 1 billion US dollars (refer to Table 9.3).

(4) In terms of investment subjects, there were large numbers of investing enterprises, with major enterprises playing an outstanding role. State-owned enterprises and limited liability companies are the main force in the outward direct investment of China, with central enterprises and entities obviously at a dominating position. In the inventory at the end of 2009, state-owned enterprises account for 69.2 %, limited liability companies for 22 %, stock companies for 5.6 %, stock cooperative enterprises for 1 %, private-run enterprises for 1 %, enterprises with foreign investment for 0.5 %, collective enterprises for 0.3 %, enterprises with investment from Taiwan, Hong Kong and Macau for 0.1 %, and others for 0.3 %. In the non-financial outward direct investment inventory, central enterprises and entities account for 80.2 %, and local enterprises for 19.8 %.

In the outward direct investment inventory of all provinces, autonomous regions and municipalities, Guangdong Province ranked at the top, followed by Beijing, Shanghai, Zhejiang, Shandong, Jiangsu, Hunan, Fujian, Liaoning and Heilongjiang (refer to Table 9.4).

Table 9.2 Countries (regions) each with OFDI flow from China over 100 million USD in 2009

S/N	Country/region	Amount (10,000 USD)	S/N	Country/region	Amount (10,000 USD)
1	China Hong Kong	3,560,057	17	Indonesia	22,609
2	Cayman Islands	536,630	18	Cambodia	21,583
3	Australia	243,643	19	Laos	20,324
4	Luxembourg	227,049	20	United Kingdom	19,217
5	British Virgin Islands	161,205	21	Germany	17,921
6	Singapore	141,425	22	Nigeria	17,186
7	United States	90,874	23	Kyrgyzstan	13,691
8	Canada	61,313	24	Egypt	13,386
9	China Macau	45,634	25	Iran	12,483
10	Myanmar	37,670	26	Turkmenistan	11,968
11	Russian Federation	34,822	27	Brazil	11,627
12	Turkey	29,326	28	Venezuela	11,572
13	Mongolia	27,654	29	Viet Nam	11,239
14	Korea	26,512	30	Zambia	11,180
15	Algeria	22,876	31	Netherlands	10,145
16	Democratic Republic of Congo	22,716		Total	5,495,537

Source: *2009 Statistical Bulletin of China's Outward Foreign Direct Investment*

Table 9.3 Top 20 countries (regions) with OFDI inventory from China in 2009

S/N	Country/region	Inventory (100m USD)	S/N	Country/region	Inventory (100m USD)
1	China Hong Kong	1644.99	11	Canada	16.70
2	British Virgin Islands	150.61	12	Kazakhstan	15.16
3	Cayman Islands	135.77	13	Pakistan	14.58
4	Australia	58.63	14	Mongolia	12.42
5	Singapore	48.57	15	Korea	12.18
6	United States	33.38	16	Germany	10.82
7	Luxembourg	24.84	17	United Kingdom	10.28
8	South Africa	23.07	18	Nigeria	10.26
9	Russian Federation	22.20	19	Myanmar	9.30
10	China Macau	18.37	20	Zambia	8.44

Source: *2009 Statistical Bulletin of China's Outward Foreign Direct Investment*

Table 9.4 Top 10 provinces, autonomous regions and municipalities with OFDI inventory at the end of 2009

S/N	Name of province, autonomous region and municipality	Inventory (100m USD)	S/N	Name of province, autonomous region and municipality	Inventory (100m USD)
1	Guangdong province	95.45	6	Jiangsu province	24.99
2	Beijing municipality	37.59	7	Hunan province	20.48
3	Shanghai municipality	35.90	8	Fujian province	15.88
4	Zhejiang province	29.59	9	Liaoning province	14.92
5	Shandong province	26.23	10	Heilongjiang province	10.62

Source: *2009 Statistical Bulletin of China's Outward Foreign Direct Investment*

The basic situation of outward direct investment of China mainly demonstrates the features of full coverage of industries, wide distribution and large number of subjects, indicating that the outward investment strategy of China complies with national conditions and has given proper play to the comprehensive advantages of Large Country. In the selection of industries, the commercial service industry, whole-sale and retail industry and mining industry account for high proportions, embodying the comparative advantage of industrial development in China; in the selection of regions, a very big proportion has been invested in Asia and Hong Kong of China, embodying the comparative advantage of similar traditional customs and consuming habits; in the selection of enterprises, state-owned enterprises, central enterprises and developed coastal regions take a very large portion, embodying the competitive advantage of these enterprises with large scale and fairly high technological level and management level.

However, the current development situation only demonstrates the features of outward investment in the primary stage of development, to raise the outward investment level, active efforts should be made to push the transformation from comparative advantage to competitive advantage, so that the comprehensive advantages of Large Country can be brought into play at a higher level. Our concrete suggestions are: first, in industrial development, the proportion of manufacturing industry in outward investment should be gradually increased, to fully demonstrate the advantage of manufacturing industry of China; the proportion of information industry and high and new technology industries should be gradually increased, to push the outward invested industries by China towards the high end of the value chain. Second, in geographic regions, the proportion in developed countries should be increased step by step, to obtain more capital and technological resources in developed countries; the early-mover advantage of some industries in China on technological level should be brought into play, to catch up the cutting edge of international science and technology. Third, in the selection of enterprises, the proportion of privately run enterprises should be increased step by step, so that they can grow quickly in the international competition environment; privately run enterprises should be encouraged and supported to "go global", and actions should be taken to upgrade the technological level and management level of privately run enterprises.

Chapter 10
Conclusions

In the research of the economic growth miracle of China, I chose the subject of Large Country economy, and made it my long-term research orientation, to make in-depth researches on the economic development law of Large Countries and the issue of comprehensive advantages of Large Country. It was found in this research that the economic competitiveness of Large Countries is mainly originated from the "comprehensive advantage of Large Country", which is a special advantage derived from the scale, differences, diversity and relative independence of Large Country economy. On this basis, the author revealed the economic development advantages of scale effect, complementarity, adaptability and relative stability derived from the features of Large Country economy in scale, difference, diversity and relative independence, analyzed the framework of comprehensive advantages of Large Country on aspects of natural resources, human capital, financial capital, technological progress, market potentials and foreign trade, and expounded the economic development strategy orientation based on comprehensive advantages of Large Country. In this way, a fairly systematic understanding has been acquired on the comprehensive advantages of Large Country, and it can also be said as the rudiment of the theory of comprehensive advantages of Large Country.

It is delightful that the rapid rise of the BRIC in the world today has made the phenomenon of Large Country economy more attractive to all people, as a new phenomenon in international economy. However, one question is always tangling on me: in the 20th century, there were the "Four Asian Tigers", which seemingly proved the advantages of small country economy; but in the 21st century, the BRIC rose, which seemingly proved the advantages of Large Country economy. Then what is really better, "big" or "small", and how can we see the objective necessity in the development of things through the complicated accident phenomena? How should we analyze the laws hidden in them in the stages of economic development? We can conjecture that Large Country economy can achieve rapid and sustained growth only when it has developed to a certain stage and has acquired some conditions. Then, what are these conditions and influencing factors? This is a question worth our analysis and judgment in a scientific way. Also, the author

Y. Ouyang, *The Development of BRIC and the Large Country Advantage*,
DOI 10.1007/978-981-10-0633-3_10

always believes that this cannot be clearly explained with the principle of comparative advantage or later-mover advantage, and only the principle of comprehensive advantages of Large Country can explain it. Therefore, in-depth theoretical and empirical researches on this issue should be helpful to the prosperity and development of economic science.

After the international financial crisis triggered by the subprime crisis in the United States, the American scholar Farad believed that, the global power is transferring, and the world is gradually getting rid of the economic governance by the United States, marching towards a post-American era supported by multiple forces. This era will see the rise of emerging Large Countries, and the United States and emerging Large Countries will jointly support the development of world economy. How should we conduct in-depth research on the characteristics and laws of Large Country economy, and make use of the Large Country effect to push forward economic growth and form comprehensive advantages favoring economic growth, and formulate scientific strategic orientation to passionately greet the era of Large Country economy rising? This is a major subject before the economists.

The BRIC have the advantages of Large Countries, and also carry the responsibilities of Large Countries. They represent the interests of emerging market countries and developing countries, and have become an important force in the international society. To greet the era of flourishing Large Country economy, the BRIC should bring into full play their advantages as Large Countries, take an active part in establishing a new international financial system, and promote the sustainable development of world economy.

First, promote the restoration of global economic growth. The BRIC have made major contributions to the world economic growth, and in the current background of financial crisis, they should cooperate better jointly to push forward the restoration of world economic growth. A report recently published by Goldman predicted that, China will restore the long-term trend growth rate of economy in 2010, India and Brazil will restore the trend growth rate in 2011 and Russia will restore the trend growth rate in 2012, all getting rid of difficulties before the developed countries. Emerging market countries are making active efforts in remedying global economy, and all governments are implementing programs to promote steady economic development, including proactive financial policies and moderately easy monetary policy, to expand domestic demand and revive the economy more quickly. The Chinese Government invested 4 trillion yuan in two years to support the development of major industries, so that the "Chinese factors" can play an important role on the international bulk commodity market; the strong growth of domestic demands in the BRIC will become the pushing force to the export-driven revival of developed economies in the coming years.

Second, pushing forward the reform of international financial system. Establishing a fair, impartial, tolerating and orderly new international financial order can provide institutional and mechanism guarantee for the sustained development of world economy, and is in line with the fundamental interests of emerging market countries and developing countries. The BRIC countries, with rapid increasing economic aggregate and foreign exchange reserves, play an important role in the international financial field. To cope with international financial crisis, all countries took an active part in the trade financing program of the International Finance Corporation, and supported capital increase to the IMF. Therefore, the influencing power of Large Countries should be brought into play in international affairs, active efforts should be made to complete the international financial supervision mechanism so that they can effectively participate in regulatory organizations such as Financial Stability Board, and emerging market countries and developing countries can have more right to speak and representativeness. In the meantime, we should speed up the completion of international monetary system, improve the reserve money issuance regulation and control mechanism, maintain the exchange rate of main reserve moneys relatively stable, and steadily push ahead the diversification and rationality of international monetary system.

Third, pushing the sustainable development of world economy. The BRIC countries have a common task in coping with international financial crisis, that is, to well combine expanding domestic demand with restructuring, and make efforts to change the economic growth pattern, to set an example in realizing sustainable development of economy in developing countries. BRIC countries are developing countries or countries in transition, still with many problems in economic system, economic structure, growth pattern and development quality, and the effective solving of these problems is critical to realizing sustainable development of economy. Therefore, effective policy orientation should be adopted according to the structural problems existing in the economic development in various countries, emphasis should be laid on supporting the high and new technology industries, modern service industry and environmental protection industry, and low carbon economy and cyclic economy be actively developed; the sustainable development of economy should be guaranteed on strategic pattern and institutional mechanism by aiming at transformation and innovation of institutional mechanism, to provide good reference to emerging market countries and developing countries.

The development economists Zhang Peigang once said that, research on the development of Large Country economy has far-reaching significance, the development economics must focus on the research on Large Countries with especially complicated problems, to reveal the laws of economic development in Large Countries and find the solutions to problems, and only by so doing that its theoretical views and policy propositions can be more universal and practical. By Large Country advantage, it mainly refers to the characteristics of Large Countries with vast territory, large population, rich resources and huge market and the economic development advantages derived from them. For example, a Large Country has a large scale of economy, usually it is easy for it to foster pillar industries, to gain advantages in specialized division of labor, an industry of ordinary scale in a Large Country can probably exceed a key industry in a small country, and this is the Large Country effect on economic growth; a Large Country has a very big domestic market, so it is usually easy to establish internal and external circulating systems for economy, to make use of both domestic and overseas markets, the domestic market can be better used by expanding domestic demand in case of turbulence on international market, and this is the Large Country effect on market stability; a Large

Country has very large aggregate saving, which enables accumulating huge amount of capital, therefore it has a strong capital regulation capacity, able to mitigate the impact from an international financial crisis, and this is the Large Country effect on financial stability. In world economic development, Large Country economy has demonstrated its unique advantages. For example in 2008, to deal with world financial crisis, the Chinese Government formulated a package program to stimulate economic growth: substantially increasing financial input, readjusting and rejuvenating industries, pushing forward technical transformation and maintaining a stable finance. With the capital advantage of a Large Country, China invested 4 trillion yuan to support the development of major industries and expand domestic demand; with the market advantage of a Large Country, China launched and opened up domestic consuming market, especially, active efforts were made to open up rural consuming market, to shift the emphasis to the domestic market in the situation of downturn of overseas market, and seek for space to expand domestic demand within the country, so as to maintain the prosperity and stability of economy by opening up domestic consuming market.

Of course, the Chinese Government can make good use of the Large Country advantages also thanks to the economic system of socialism with Chinese characteristics, which enables better use of both means of planning and market to regulate economy. Modern economics has told us that, market is an invisible hand, while planning is a visible hand. Adam Smith praised highly the "invisible hand", believing that market mechanism can spontaneously regulate social economy, to coordinate all links in production and maintain balance between supply and demand; Keynes praised highly the "visible hand", stating that the spontaneous regulation by market mechanism cannot realize balanced development of social economy, therefore economic intervention and regulation by the state is required. When total demand is insufficient, mainly the state will attempt to stimulate and expand total demand, with the focus on investment, so that the national income can increase in multiples following the multiplier principle, to increase consumption demand and realize steady development of economy. In the recent two years, due to the impact from international financial crisis, insufficient demand and economic downturn appeared in China. To cope with the financial crisis, the government adopted both means of market and planning, especially the "visible hand", and proactive financial policies, monetary policies and industrial policies were implemented, with fairly good results.

We maintained economic stability by relying upon the Large Country effect, and should make better use of this effect to promote economic development. China regulated and optimized its industrial structure, advanced technical transformation and strengthened the construction of infrastructure during the period of international financial crisis, to create better conditions for improving quality and efficiency in economic development and realizing better and faster development. It can be predicted that, by transforming the development pattern and upgrading industrial structure, the Chinese economy will perform better in the world after the crisis.

References

An'gang H (1998) Make full use of advantages of large countries, and actively expand domestic demand. Research Report on National Conditions of China 24

Bai M (2008) Rise of China and India calls for innovation of large country development pattern. Asia-Pacific Econ 1

Barrow RJ et al (2000) Determinants of economic growth. China Social Sciences Publishing House, 2000 Edition

Bei J (2001) On the properties of enterprise competitiveness. China Ind Econ 10

Bei J (2003) Competitiveness economics. Guangdong Economic Press, 2003 Edition

Bi Q, Bao Y, Li B (2007) Correlation analysis of population delivery and regional economy coupling in inner Mongolia. Geogr Res 5

Cai F, Du Y (2000) Convergence and differences in regional economic growth in China—enlightenment to the strategy of developing the west. Econ Res 10

Central Compilation and Translation Bureau of Works of Marx, Engels, Lenin and Stalin (1997) Complete works of Marx and Engels. People's Publishing House, 1997 Edition

Chandler A Jr (2006) Scale and scope: the dynamics of industrial capitalism. Huaxia Publishing House, 2006 Edition

Chen W (1994a) Breakthroughs in development economics and puzzles in large country development. Jianghan Tribune 10

Chen W (1994b) Ten major puzzles in the development of large countries—exploration on difficult points in large country development economics. Hubei People's Publishing House, 1994 Edition

Chen B, Feiqiong R (2008) Selection of China's outward direct foreign investment. Stat Res 8

Chen X, Zhang R (2006) Research on contribution of heterogeneous human capital in regional economic gap, Econ Perspect 3

Chenery HB, Syrquin M (1988): *The Patterns of Development*, Economic Science Press 1988 Edition

Chenery et al (1995) Industrialization and growth: a comparative study. Sanlian Bookstore and Shanghai People's Publishing House, 1995 Edition

Cheng H (1998) Research on comparative advantage of outward direct investment. Shanghai Sanlian Bookstore, 1998 Edition

Cheng S (2001a) Strategy and management of China's overseas investment. Democracy and Construction Publishing House, 2001 Edition

Cheng X (2001b) Research on regional comparative advantage in china. China Planning Publishing House, 2001 Edition

Cheng H (2004) Outward direct investment development strategy of privately run enterprises of China. China Social Sciences Publishing House, 2004 Edition

Cheng J, Li J (2006) Comparative research of economic and social development in five large countries (1990–2005). Economic Science Press, 2006 Edition

Y. Ouyang, *The Development of BRIC and the Large Country Advantage*,
DOI 10.1007/978-981-10-0633-3

China Modernization Strategy Research Subject Group (2004) Regional modernization in India. China Net Feb 20.
Cui Y (2005) Double-edged sword of trade and development, Reading 11
Dai Q, Bie Z (2006) FDI, Human capital accumulation and economic growth. Econ Res 4
Daokui L (2006) China's economy needs large country development strategy. China Today Forum 7
Daokui L (2007) Large country development strategy. Peking University Press, 2007 Edition
Deming L (1999) Motivation analysis on economic development of China. Shanxi Economics Publishing House, 1999 Edition
Diankai W (2002) Comparative advantage trap and china's choice of trade strategy. Econ Rev 2
Ding D, Zhibiao L (1999) From human capital to heterogeneous human capital. Prod Res 3
Donghui S (1999) Introduction to industrialization in late-mover countries: a survey of history of industrialization and industrialization history. Shanghai University of Finance and Economics Publishing House, 1999 Edition
Duan P (2007) Causes of unbalanced domestic and overseas economy of China and countermeasures. East China Econ Manag 9
Fan Z (2004) Research on transnational operation by small and medium-sized enterprises of China. China Social Sciences Publishing House, 2004 Edition
Fengde J (1988) General characteristics of economic big powers and Japan as a economic big power. Japanese Studies 5
Gao F (2007) Challenges faced by China towards large country economy. China National Conditions and Strength 3
Gaolan H (2005) Capital accumulation and economic catch-up, Contemp Econ Res 11
Ge L et al (1999) Comparative study of economic growth patterns of main countries (regions) in the world. Acad Mon 5
Gillis et al (1998) Economics of development. China Renmin University Press, 1998 Edition
Gong Y (2001) Collection of theories on large country economy development. Jianghan Tribune 9
Guan H (2007) Effectiveness of comparative advantage theory: inspection based on historical data of China. Econ Res 10
Guoguang L (2009) New theories of economics. Social Sciences Academic Press, 2009 Edition
Guohua P (2007) Total factor productivity and human capital composition in China. China Ind Econ 2
Guomin L, Wang Y (2003) On relations between comparative advantage and viability. Econ Res 9
Guoyuan G (2008) Selection of industries for china's outward direct investment. Stat Res 9
Hai W (2007) China's economy has advantages of large countries and can grow powerfully for further 20 years. Beijing Morning Post June 1, 2007
Haibing G (2005) Large country and large country economy development strategy. Pac J 1
Heyne P et al (2008) The economic way of thinking. World Publishing Corporation, 2008 Edition
Hou J, Jiang Y (2004) Comparative advantage and catch-up strategy. J Pub Manag 11
Hu D et al (2002) Economics in a real world—a survey of neoinstitutional economics. Contemporary China Publishing House, 2002 Edition
Huaguang Y (2009) Extension of theories and strategic adjustment for china's outward direct investment. Chinese and Foreign Entrepreneurs 2
Huang L (2002) Economic recession in the United States: causes and its impact on world economy. Internal Circulation 1
Jiagui C, Qunhui H (2005) Development of industry, changes in national conditions and economic modernization strategy—analysis of national conditions for China becoming a big industrial country. Soc Sci China 4
Jing X (2000) Economic development patterns of large countries and main supporting points to economic growth in China. Shanghai Econ Rev 5
Jinyong L (2005) Free trade areas: China's large country economy, Business Week 7
Kennedy P (2006) The rise and fall of the great powers. International Culture Publishing House, 2006 Edition

Kesha G (2003) Comments on foreign trade strategy and trade policies of China. Int Econ Rev 1
Kesha G (2004) Selection of china's industry development strategy and policies. Soc Sci China 1
Kunrong S (2003) New growth theory and economic growth in China. Nanjing University Press, 2003 Edition
Kunrong S, Ma J (2002) The "club convergence" characteristic in economic growth in China and analysis of its origin. Econ Res 1
Kunsu L, Chen L (2005) Empirical analysis of economic growth convergence in APEC. World Econ 9
Kuznets S (1985) Economic growth of nations. The Commercial Press, 1985 Edition
Kuznets S (1989) Modern economic growth. Capital University of Economics and Business Press, 1989 Edition
Lewis A (2002) The theory of economic growth. The Commercial Press, 2002 Edition
Li Z (1999) Human capital—a theoretical framework and its explanation to some issues of China. Economic Science Press, 1999 Edition
Li X (2000a) Scientific selection of regional economy development strategy. Truth Seeking 4
Li Y (2000b) On large country economy. Beijing Normal University Press, 2000 Edition
Li Y (2009) Research on outward direct investment results of chinese enterprises. Manag World 9
Lin JY (2002a) Development strategy, viability and economic convergence, Econ (Q.) 2
Lin JY (2002b) Reflection on viability, economic transformation and new classical economics. Econ Res 12
Lin YJ, Cai F, Li Z (1999) China's miracle: development strategy and economic reform (revised and enlarged edition), Shanghai Sanlian Bookstore and Shanghai People's Publishing House, 1999 Edition
Lin JY, Cai F, Zhou L (1998) Analysis of regional gap in economic transformation period of China. Econ Res 6
Lin JY, Cai F, Zhou L (1999) Comparative advantage and development strategy—reinterpretation of "East Asia Miracle". Soc Sci China 5
Lin JY, Li Y (2003) Comparative advantage, competitive advantage and economic development in developing countries. Manag World 7
Lin JY, Liu M (2003) Economic development strategy and China's industrialization. Econ Res 7
Lin JY, Peiling L (2003) Economic development strategy and regional income gap in China. Econ Res 3
Lin JY, Shiyuan P, Liu M (2006) Selection of technologies, institutional system and economic development. Econ (Q.) 3
Lin JY, Xifang S (2003) Comparative advantage strategy theory for economic development—also on "Comments on China's foreign trade strategy and trade policies". Int Econ Rev 11–12.
Liu X (2004) Analysis of result of privatization restructuring on the industrial efficiency in China, Econ Res 8
Liu C (2009) Problems of chinese enterprise in outward investment and countermeasures. Pioneering with Sci Tech Mon 5
Liu Z, Hu Y (2008) Empirical test of total factor productivity, capital accumulation and regional differences. Statistics and Decision-making 12
Liu S et al (2004) Gray system theory and its applications, Science Press, 2004 Edition
Liu Y et al (2006) Correlation analysis of coupling of regional urbanization with ecological environment in China. Acta Geographica Sinica 2
Lu T (2003) Transnational operation strategy of chinese enterprises. Economic Management Press, 2003 Edition
Lu M et al (2008) Reform and opening-up 30 years: large country economy development road of china (economy volume), Encyclopedia of China Publishing House, 2008 Edition
Lynn SR (2009) Development economics. Truth & Wisdom Press, Shanghai Sanlian Bookstore and Shanghai People's Publishing House, 2009 Edition
Marshall A (1964) Principles of economics. The Commercial Press, 1964 Edition

Ouyang Y (2005) Analysis of advantages of transnational operation by privately run enterprises of China. Manag World 5
Ouyang Y (2006a) Build up comprehensive advantages of large country, J Hunan University of Commerce 1
Ouyang Y (2006b) The direct outward investment strategy of china based on comprehensive advantages of large country. Econ Financ Trade 5
Ouyang Y (2007) Outward direct investment of privately run enterprises: theory, strategy and pattern. The Commercial Press, 2007 Edition
Ouyang Y (2009a) Proposing of comprehensive advantages of large country and research approach. Econ Perspect 8
Ouyang Y (ed) (2009b) Research on large country economy (the first series). Economic Science Press, 2009 Edition
Ouyang Y (2010a) Large country effect of the rise of BRIC. Guangming Daily Feb. 2
Ouyang Y (ed) (2010b) Research on large country economy (the second series). Economic Science Press, 2010 Edition
Ouyang Y et al (2009a) Comprehensive advantages of large country: a new annotation of China's economy competitiveness. Econ Theory Econ Manag 11
Ouyang Y et al (2009b) Outline of comprehensive advantages of large country. J Hunan University of Commerce 1
Ouyang Y, Sheng Y (2008) Technological gap, technology adaptive capacity and late-mover technological catch-up. China Soft Sci 2
Ouyang Y, Sheng Y (2010) Diversified technologies, adaptive capacity and coordinated development of regional economy in late developing large countries. Econ Rev 4
Peigang Z (1992) New development economics. Henan People's Publishing House, 1992 Edition
Pilny K (2008) How India and China change the world. International Culture Publishing House, 2008 Edition
Porter M (2002) The competitive advantage of nations. Huaxia Publishing House, 2002 Edition
Qi H (2006) Theoretical exploration on development pattern of developing large countries—analysis based on labor force market. Theor Circle 5
Qian X et al (1981) Principles of dialectical materialism. People's Publishing House, 1981 Edition
Qian X et al (1983) Principles of historical materialism. People's Publishing House, 1983 Edition
Ren B, Liu L (2008) 30 years of "economic growth miracle" in china: description, definition and theoretical interpretation. J Northwest University (Philos Soc Sci Edn) 1
Research Group of Fudan University World Economy Research Institute (2006) Institutional system transition and restructuring: large country economy development track since the 1990s. Shanxi Economy Publishing House, 2006 Edition
Schumpeter J (1990) Theory of economic development. The Commercial Press, 1990 Edition
Schweitzer GE (2007) The science and technology, economy and security in Russia in the new period. Beijing Institute of Technology Publishing House, 2007 Edition
Shanyong L, Yan W (2005) A rustic opinion on comprehensive advantage development strategy—new thinking on economic development beyond the debate of comparative advantage strategy theory and catch-up strategy theory. Reform and Strategy 8
Shen K, Quan H et al (2008) Comparative study of economic development patterns—comparison of economic development in China and India. SASS Publishing House, 2008 Edition
Smith A (1972) An inquiry into the nature and causes of the wealth of nations. The Commercial Press, 1972 Edition
Song Y (2001) Research on transnational direct investment by Chinese enterprises. Dongbei University of Finance and Economics Press, 2001 Edition
Song Y, Fang J (2007) Research on correlation of economic fluctuation in China and the world. Econ Financ Trade 1
Song Z, Kang F (1994) World economic history (vol I). Economic Science Press, 1994 Edition

State Statistics Bureau (1978–2009) Statistical bulletins of national economy and social development of the People's Republic of China 1978–2009
State Statistics Bureau (2010) China Statistical Yearbook (2010). China Statistics Press, 2010 Edition
Sun J (2000a) Internationalized operation of capital. Economic Science Press, 2000 Edition
Sun Y (2000b) Manuscript of socialist economic theory. Guangdong Economic Press, 2000 Edition
Sun L, Ruoen R (2005) Estimation of capital input and total factor productivity in China. World Econ 12
Tang J (ed) (2006) Rise of large countries. People's Publishing House, 2006 Edition
Tao J (2005) A view of China's later-mover advantage from explicit advantage. World Econ Stud 1
The World Bank (2009) World development report, Tsinghua University Press, 2009 Edition
Tian Q (2001) Comparison of internationalization pattern of large country economy and small country economy. Int Econ Trade Res 1
Wang J (2001) Human capital and economic growth: theory and demonstration. China Finance Publishing House, 2001 Edition
Wang Y (2002) WTO and China's trade development strategy. Economic Management Press, 2002 Edition
Wang L (2003) Transnational operation theory and strategy. University of International Business and Economics Press, 2003 Edition
Wang Q (2004) China's outward direct investment: theory, practice and recommendations. Int Econ Cooperation 3
Wang S (2007) Theory and empirical study on comprehensive comparative advantage. China Social Sciences Publishing House, 2007 Edition
Wang R (2008a) Comparative study of economic development in the United States and Brazil. Economic Science Press, 2008 Edition
Wang J (2008b) Transformation and economic growth—research based on solow model. Fudan University Press, 2008 Edition
Wei X (2001) Economic consideration on the history and reality of economic development in China. Econ Res 7
Wei H (2004) Comparative advantage, competitive advantage and regional development strategy. Fujian Forum 9
Weil DN (2008) Economic growth. China Renmin University Press, 2008 Edition
Xiahui L, Zhang P, Zhang X (2008) Economic growth and structural transformation in era of reform. Truth & Wisdom Press and Shanghai People's Publishing House, 2008 Edition
Xiaojuan J (2004) Attracting foreign investment, outward investment and China's goal of all-round well-off society. J Int Trade 1
Xiaojuan J, Yang S, Feng L (2003) China foreign economic relations and trade theory front III. Social Sciences Academic Press, 2003 Edition
Xibao G, Hu H (2003) Later-mover advantage strategy and comparative advantage strategy. Jianghan Tribune 9
Xibao G, Hu H (2004) New theory of later-mover advantage—also on the motive force of economic development in China. J Wuhan University 3
Xing J (2003) Outward direct investment: strategic selection. Economic Science Press, 2003 Edition
Xinhua J, Zhang H (2007) On transformation of foreign trade growth pattern of China. China Ind Econ 8
Xu D (2002) "Going Global": status quo, problems and countermeasures. Int Econ Cooperation 4
Xu Y (2003) Research on unsuitability of comparative advantage strategy in economic development of China. Reform 5
Xu C (2009) Policies supporting the development of modern agriculture in Brazil. Mod Agric Sci Technol 6

Xu X, Ma Y (2007) Primary exploration on large country economy strategy of China. Heilongjiang Foreign Econ Relat Trade 3
Xu J, Yin X (2002) Deterioration of trade conditions and effectiveness of comparative advantage strategy. World Econ 1
Yang S (2007) Selection of economic growth route for China. Econ Perspect 2
Yang R, Yang Y (2007) Limited catch-up and development of large country economy, Essay at Sixth Annual Conference of China Economics in 2007
Yang X, Zhang Y (2001) New trade theory, comparative interest theory and new results of their empirical research: review of literatures. Econ (Q.) 1
Yinxing H (1997) From comparative advantage to competitive advantage—concurrent discussion on defects of the theory of comparative advantage of international trade. Econ Res 6
Youhao T (1999) Superficial theory of large country economy—also on economic development strategy of China. Econ Syst Reform 3
Youhao T (2001) Research on large country economy and open economy. Pac J 2
Yu W, Botao Q (2006) Large countries and comparative advantage development strategy. Prediction 5
Yu J, Kejian G (2005) Change of factor supply advantage and transformation of industrial and trade structure under opening up conditions. J Int Trade 11
Yuan Z, Jianyong F (2003) Analysis of industrialization process in China since 1978 and its regional differences. Manage World 3
Yun J (2003) On composite comparative advantage. World Econ Study 6
Yunshi M (2001) Strategic competition of transnational companies and international direct investment, Zhongshan University Press, 2001 Edition
Zakaria F (2009) The Post-American world. China CITIC Publishing House, 2009 Edition
Zhang P (1992) New development economics. Henan People's Publishing House, 1992 Edition
Zhang H (2001) Large country pattern and economic growth in China. Econ Theo Econ Manage 9
Zhang L (2007) Advantages of large countries and economic growth potential of China. Mod Econ 6
Zhang X, Jin X, Ningning Z (2006) Trade opening up, preference on foreign direct investment and regional economic growth in China—view based on large country economy. J Int Trade 12
Zhang P, Xiahui L (2007) Front of economic growth in China. Social Sciences Academic Press, 2007 Edition
Zhang X, Xiaozhong L (2001) Empirical analysis of comparative advantage of foreign trade products of China. J Quant Techn Econ 12.
Zhang Y, Xianzhong Y (2007) Circling and layering upgrading of trade structure and change of foreign trade growth pattern in heterogeneous large countries. Int Trade 2
Zheng J (2007) How to define large countries. Stat Res 10
Zheng H, Ruoen R (2005) Research on international competitiveness of the manufacturing industry of China under multilateral comparison: 1980–2004. Econ Res 12
Zhou Y (2004) Ownership advantage of privately run enterprises of china in implementing the "going global" strategy. J Univ Int Bus Econ 4.
Zhuoyuan Z (ed) (2000) Economic consideration of china's experience on reform and opening up. Economic Management Press, 2000 Edition
Zong J (2007) Effect of world large country economy fluctuation on china's economy—empirical analysis based on the view angle of FDI with G7 as example. Forum of World Econ Politi 4
Zou W, Dai Q (2003) Technological imitation, human capital accumulation and economic catch-up. Social Sciences in China 5

Postscript

In the mid 1980s, I studied for the degree of Master of Philosophy in Renmin University of China. Although we studied philosophy, the monumental works of Karl Marx on economics *On the Capital* was a compulsory course. This classical works on the social forms of capitalism contains extensive and profound knowledge with stringent and precise logic thinking. As precious deposits of thinking, we learned from it the thinking method and ideas of dialectical materialism and historical materialism, as well as the economic theories of Marxism. I became interested in the economics since then, and that's why I chose to study for the Doctor of Economics later.

As the human society came to the 21st century, with the daily-changing progress of sciences and technologies, the pace of social and economic development has accelerated and the trend of economic globalization become more apparent; in the course of world economic development, the economic take-off of the "Four Asian Tigers" was followed by the miracle of rapid economic growth in China, accompanied with the rise of the BRIC countries, therefore, the economic rise of Large Countries has become an international phenomenon. I made initial exploration on the issue of Large Country economy when studying for the doctor degree in Hunan University, putting forth the concept of "comprehensive advantages of Large Country" and analyzing the economic characteristics and development strategies of late-developing Large Countries. In autumn 2006, I came to the postdoctoral research station on practical economics of the Institute of Fiscal Science of the Ministry of Finance, continuing my research on Large Country economy and comprehensive advantages of Large Country, and was assigned the project *Formation Mechanism and Framework Analysis for Comprehensive Advantages of Large Country* specially financed by China Postdoctoral Science Foundation. The superior academic environment in this institute provided me with a new platform for my study and research. My cooperation supervisor Prof. Liu Shangxi is a young and promising scholar, and I benefited a lot from him with his exuberant knowledge in economics, rich practical experience, rigorous scholarship, and broad thinking space. Prof. Jia Kang, Prof. Su Ming and Prof. Wang Zhaocai in the Institute provided highly valuable opinions, and helped me to improve my postdoctoral report; the division head Ji Ziying of China Postdoctoral Management Council and

Y. Ouyang, *The Development of BRIC and the Large Country Advantage*,
DOI 10.1007/978-981-10-0633-3

division head Zhang Yeping of the Institute of Fiscal Science of the Ministry of Finance offered their cordial assistance to my research. Dr. Sheng Yanchao, Dr. Yi Xianzhong and Dr. Liu Zhiyong, and Lecturer Luo Huihua of the Large Country Economy Research Group of Hunan University of Commerce collected research literatures for me, conducted exchanges and discussions with me, and also cooperated with me in some academic papers. Therefore, my postdoctoral report also contains the hard work of my teachers and colleagues.

The "Contemporary Economics Series" compiled with Mr. Chen Xin as the chief editor presents the latest research results in the contemporary economics, enjoying high reputation in the academic circle of China. Therefore, when I started my study for the doctor degree, I hoped that my works can be included in the "library of contemporary economics" of this series. I thank from the bottom of my heart the editor-in-chief Chen Xin and He Yuanlong, President of Truth and Wisdom Press, who have realized my dream. To improve the quality of the manuscript of the book, I made major modifications to it on the basis of my postdoctoral report, so that the book is more logic in the organization of ideas and more rational in structure.

My teachers and friends in the academic circle, especially Prof. Liu Guoguang, Prof. Zhang Zhuoyuan, Prof. Yang Shengming, Prof. Liu Shucheng, Prof. Zheng Yuxin and Prof. Wei Houkai of Chinese Academy of Social Sciences, Prof. Fang Fuqian of China Renmin University, Prof. Cai Jiming of Tsinghua University, Prof. Zhang Yabin of Hunan University, Prof. Liu Siwei of Hunan University of Commerce, as well as Prof. Dmitry Sorokin, corresponding member of Russian Academy of Sciences, Prof. Scott Rozelle of Stanford University USA and Prof. Santosh C.P. Dean of the School of Economics of Delhi University of India provided beneficial proposals to my research. I would like to extend my sincere thanks to all of them together here!

Yao Ouyang
Mengze Garden, Changsha, Jan. 2011

Milton Keynes UK
Ingram Content Group UK Ltd.
UKHW021959181223
434618UK00002B/49

9 789811 092152